Essential Career Skills For Engineers

By

Shahab Saeed, P.E.

Keith Johnson, P.E.

Engineering & Management Press
Institute of Industrial Engineers
Norcross, Georgia, U.S.A.

00 99 98 97 96 95   6 5 4 3 2 1

Library of Congress Cataloging-in-Publication Data
Saeed, Shahab, 1959-
    Essential career skills for engineers / by Shahab Saeed, Keith Johnson.
      p. cm. -- (Engineers in business; 1)
    Includes bibliographical references and index.
    ISBN 0-89806-142-3 (pbk.) : $25.00 ($20.00 IIE members)
    1. Engineering -- management. 2. Problem solving. 3. Group problem
solving. I. Johnson, Keith, 1955-   . II. Title. III. Series.
TA190.S24   1994
620' .0023--dc20   94-36387
                                                          CIP

Director of Publications: Cliff Cary
Acquisitions Administrator: Eric E. Torrey
Book Editor: Forsyth Alexander
Design: Jeannie Glover, Kim Knox Norman,
             Steven Prazak, Patrick Worley

ISBN 0-89806-142-3

Engineering & Management Press
25 Technology Park
Norcross, GA 30092
770/449-0461 phone
770/263-8532 fax

Table of Contents

Acknowledgments

For his extraordinary efforts in reviewing the manuscript and providing insightful improvements, our special thanks goes to Steve Stepanek, president of a regional energy company, and a successful engineer in his own right. We also thank Linda Aghdassi, Borzu Sohrab, and Shahrokh Saeed for their willing participation and helpful suggestions. A special thanks to Joe Logan for sharing his wisdom and passion for empowerment. Without the consideration, constant encouragement, persistence, and patience of our editor, Maura Reeves, we could not have completed the work. We must also recognize the important contribution of countless friends, colleagues, and associates, whose examples have provided the catalyst for our own learning and understanding, and whose sharing of ideas and techniques has provided much of the substance of our philosophy.

The contributions of Jan Saeed and Allison Johnson are indescribable. Even questions like, "So, when *will* it be done?" have been helpful. Lawrence Naysahn, Lauren Lahdan, Danielle Rahz, and Renae Michelle have not only lent their names to case studies but have also given soul to our work.

Preface

Ideas started flowing when we were asked to write this book. We thought about successful engineers we knew and what skills they possessed and how they used them. We reflected back to the start of our careers and considered what words of wisdom we would like to have known at that time and finally we thought about skills we look for in the engineers we hire in our company. We know what skills we would like them to have, but deciding whether a candidate possesses these skills based only on a resume and a couple of interviews is a very difficult challenge.

As you will notice from the titles of the chapters, competency in these areas is not easily discernible from interviews and resumes. However, it is possible to assess these skills. Our hope is that by perusing the concepts presented in this book, the reader will increase the probability of success in the interview process when looking for a job. These concepts should also assist engineers who are involved in the selection process of filling positions in their departments, since it is becoming increasingly common for peers to be involved in the interviewing and hiring process, even though they may not be supervisors.

The objective of this book is to stimulate thinking and raise the level of awareness about a set of skills which we believe are the common variables identified up to this point in the equation for success in industry. A mathematical representation of this formula is illustrated in Figure P.1. It seems to us that some of the "softer" issues become easier to grasp if they are expressed mathematically (at least for engineers). On the other hand, trying to express real-life situations with rigid mathematical formulae is inherently dangerous. Obviously, the items

$$Success = k_1V_1 \cdot k_2V_2 \cdot k_3V_3 \cdot k_4V_4 \cdot k_5V_5 \cdot k_6V_6 \cdot ... \cdot k_nV_n$$

Figure P.1: A mathematical representation of success

discussed in the book such as skills and personal attributes cannot be reduced to a quantitative discussion. The equation is meant to convey the relative importance of the different dimensions in which an engineer must develop skills and the fact that the relation changes over time and in various situations.

We discussed at length whether the elements of success represented in our formula should be multiplied or summed. Finally, we decided in favor of multiplication because each of the elements was considered crucial, and the total absence of any one would preclude vocational success. In other words, if any of the factors is zero, the result is also zero. To sum the elements would suggest that an unusual strength in one would compensate for total inability in another, but this is not the case. Obviously, each of us has particular strengths, but it is still necessary to be well-rounded if we want to be successful.

As the equation suggests, there are many variables that contribute to success in a career. In this book we focus on six skills which we consider essential: customer satisfaction, team dynamics, technology, finance, data analysis, and communication.

We have not attempted to quantify the value of coefficients in the model. This is left to the reader. From company to company and from industry to industry the coefficients associated with each of these variables will change and in many cases there will be additional factors that are specific to a particular job and company.

We cannot help with the coefficients and company or industry specific variables in the context of this book. However, we can help by providing a framework through which the reader can have a greater sense of what industry wants and expects from engineers in general. Then by applying the reasoning and problem solving skills they possess, the relative importance of each skill can be determined, whether from one project to the next or from one job to another. This is by no means a cookbook for success—nothing can replace sound judgment.

Another issue that came up was a suitable definition of success. We chose not to focus much on success as advancement up the management ladder, or as career progression in a technical area. For this book, we define success as the ability to make a contribution toward the goals of the organization and the quality of life within the organization. In other words, success is the degree to which one can add value. We feel this does not preclude career development, and is in fact a great contributor to advancement and perhaps more importantly to the sense of well-being that results from the knowledge that an individual is making a valued contribution.

There are seven chapters in this book, one chapter for each of the six essential skills/variables plus an introduction of the main theme and the reasons we believe the topics presented in each chapter are critical to a productive career in business today and in the years to come. As to the order of chapters, we felt that the ultimate measure of success is customer satisfaction and it should be in the forefront of every engineer's mind as well as this book. From there we have attempted to address the issues as they would be encountered in a typical project. Chapter three deals with teamwork and what the business world expects of engineers with regard to this element. In chapter four we discuss the critical nature of managing technology as a means and not an end to itself. The next chapter addresses the issue of data analysis to get a clear picture of the problem at hand and possible solutions. Chapter six is about business and the bottom line. Remember the jokes about business school being for engineering dropouts? Well, it's payback time. It is time to study how the bottom line of an enterprise is impacted by our decisions and that skills in this area are essential for a successful career. Finally, we will address the power of effective communication. It is what you say and how you say it that determines how effective you will be in your work group.

Chapter 1

The Holistic Engineer

The story is told about an engineer who worked for an earth-moving equipment manufacturer. He had worked on a project to design a new piece of equipment—let's say a bulldozer—that would match the competition in power and performance and beat it in fuel efficiency. The engineer spent long hours researching the latest technology and improving the design of the various components and was able to meet all of the project's objectives.

Proud of his accomplishments, he was surprised to hear the bulldozer was not selling well and was shocked when a salesperson told him that customers were telling her the competitor's model had better fuel economy. The engineer said, "That is impossible. I have checked and rechecked my calculations. I have reviewed and analyzed all the data. There is no way their bulldozer can have better fuel economy than ours."

The salesperson told the engineer, "Come with me to visit a customer who just placed an order with our competitor. Let's find out why they didn't choose ours so we don't lose the next sales opportunity."

He reluctantly agreed to join her on a site visit the next day. He wasn't especially excited to go. After all, he hadn't gone to engineering school for years to spend his day visiting with customers, he should be in front of his computer crunching numbers and designing sophisticated systems. Customer visits and sales "stuff" was for business school graduates, he thought.

The next day at the customer's work site, the engineer noticed this company had both of the competing bulldozers in their fleet and had been using them for a while. The salesperson explained to the supervisor in charge that they realized their company did not get the order for the additional bulldozers but wanted to spend a few minutes to see how they could improve their product to have a better chance in the future. The engineer started by asking questions about the handling of the machine, it's power, ease of operation, etc. Finally, he got to the point and asked about the fuel efficiency of his bulldozer compared to that of the competitor. The client company's supervisor said he thought the competitors model got better mileage. The engineer asked, "Why do you think so?"

The supervisor said, "The competitor's model does not need refueling before the end of one shift but yours does."

The engineer finally understood where the problem was. He was relieved that his calculations and designs were correct, but he had won the battle and lost the war! The culprit: the size of the fuel tank! The competitor's bulldozers did not have better fuel economy, just a larger tank, which created the illusion of being more efficient and swayed the customer's perception and therefore the sale!

The moral of the story is that the days are over when engineers could just sit in an office with their computer and design to their heart's delight and then throw it over the wall to manufacturing, sales, or some other internal customer. Businesses are looking for engineers who can visualize the big picture, people who see the forest *and* the trees. They are looking for holistic engineers.

We use the term "holistic" because it is one of the few words that adequately describes the notion that an engineer must combine several interdependent skills, and this combination is greater than the sum of the parts. Holistic also implies an understanding of the big picture, an appreciation for all variables and an avoidance of sub-optimization. No one is going to tell you, "We are looking for a holistic engineer," but they are.

Finding holistic engineers

Businesses are trying to find better and more accurate ways to identify these folks from among a multitude of candidates. Some use various profile tests and written exercises. In our shop we use team interviewing to increase the chances of selecting a holistic engineer. Usually the director of the department and three to four engineers are involved in this process. After we have decided how many candidates will be interviewed, each engineer investigates a particular skill with the candidate. These assignments are based on the engineer's skills and interests. For example, the individual who is the strongest in the department with regard to technology issues focuses his or her interview with the candidates on this particular topic. Another engineer builds an interview around business topics and yet another explores analytical and technical issues. Meanwhile, we all assess communication skills, level of comfort with the team approach and the degree of customer satisfaction orientation.

The candidate spends about thirty to forty-five minutes with the director of the department, and is then informed that the rest of their interview is with a number of engineers in the department in which they will be working if they are selected for the job. The results of this process have been very positive. Not only have we been successful at finding holistic engineers, we have learned a great deal about ourselves in the process. We experienced firsthand that five minds working together can uncover facts that one individual would not. This is especially true if that one individual, who is traditionally the only one conducting the interview, is the supervisor of the department.

The Lone Ranger

There was one candidate, let's call him the "Lone Ranger" for reasons that shall become apparent, who impressed the daylights out of the director with his talents and skills, but blew it with every one of the engineers. This non-holistic applicant was from the old school, where the only one he had to please was "The Boss." The Lone Ranger, during his interviews with the other engineers, made it clear that he was going to get ahead of them once he got the job. The cocky attitude earned him very low scores in teamwork from his would-be (could-have-been) teammates and resulted in his riding into the sunset without an offer. The Lone Ranger failed to realize that teamwork is a critical variable for success in our organization. He should have at least picked up on the strong clues suggesting its existence, such as having to be interviewed by peers. He probably assumed the formula for success was cast in concrete years ago, and he was not even looking to see if the multiplier for each of the variables had changed, let alone the introduction of new variables.

Mr. Rhinestone

Then there was Mr. Rhinestone, a talented, recent graduate with solid technical skills and an excellent grade point average. He scored high with everybody except one engineer. After all the interviews were completed, we got together in a conference room and compared notes, tallied the scores, and reached a consensus on who should receive the job offer. Mr. Rhinestone was one of the top three contenders until a female engineer of our team reported her concerns about this candidate and his condescending attitude towards her during the interview. None of the male engineers had sensed this attitude during their interviews. After a substantive discussion we decided that enough concern existed regarding Mr. Rhinestone's ability to deal with diversity issues to take him out of contention.

In today's global marketplace, individuals who are not sensitive to diversity issues and do not know how to utilize all the rich resources that are available, are a liability to their organization. Imagine how Mr. Rhinestone would have worked on a cross-functional task force, half of whom were women? Would he take women's ideas seriously? Would his prejudices allow him to recognize and pursue the best solution to the problem at hand? Maybe. Maybe our conclusions about Mr. Rhinestone were wrong and he was not sexist. But what if they were correct? At that point in time there were other strong candidates available and we did not want or have to take the risk. The business world is becoming more competitive each day and organizations that possess less liability and more assets, in human terms, have a much higher probability of success in the long run.

The previous examples highlight characteristics that you do not necessarily want in a new employee. There is hope, however. There are positive characteristics of a holistic engineer. These characteristics include: competency, effectiveness, and efficiency.

Technical competency

Engineers, for the most part, graduate from college with a complete set of technical skills. The challenge over the length of their career is to make sure that their arsenal of weapons is up-to-date. Competency has to do with knowing which tool to use to solve a particular problem as well as knowing how to use the tool. The greater the variety of tools in an engineer's tool box, the higher the probability of success.

A weapons analogy may help paint a clearer picture. Imagine two camps engaged in battle. Both sides are conducting bombing raids on each other's territory. One side is trying to fend off the incoming aircraft with machine guns while the other side has surface-to-air missiles at their disposal. On which side of the conflict would you prefer to be? This is not to say the use of machine guns at some point may not be appropriate. However, the point is that the wider the variety of weapon systems available to an individual, the higher the chance of victory. Abraham Maslow has said that if the only tool you have is a hammer, every problem looks like a nail.

We realize the war and battle analogy is not consistent with a holistic approach. However, the business world is very competitive and there is a battle going on for market share, customer loyalty, and the growth of the enterprise. These battles are based on intellect and ingenuity.

In today's international marketplace, businesses are looking for engineers that think globally and act locally. Technical competency is an essential prerequisite to the accomplishment of this challenging task. Obsolescence is an ever-present danger that can be mitigated by lots and lots of reading and participation in professional and industry conferences, not only as a part of the audience, but also as a presenter. Presenting to a group of peers is an experience that hones one's skills and will probably lead to new discoveries. There was an operations research professor who once told one of the authors that he understood dynamic programming the *third* time he taught it! He is still looking for the list of the students in the first three classes to issue a recall.

Professional societies can also be a great asset in this arena. They provide a forum where academia and industry can exchange information. Societies undoubtedly have as one of their goals the communication of leading edge information to their membership. Keeping up to date is critical and cannot be overstressed. The only constant in these times of rapid change seems to be the increase in the rate of change. Alvin Toffler has said that the illiterate of the twenty-first century will be those that cannot learn, unlearn, and relearn.

Effectiveness and its components

Effectiveness is another crucial characteristic of the holistic engineer. It addresses the issue of doing the "right" things. It involves attacking the root cause of the problem as opposed to its symptoms. It encompasses an understanding of the concepts of empowerment and coaching. It includes a mastery of communication challenges existing within the business world. It demands adherence to the highest standards of ethics. It requires authenticity in business relationships and not losing sight of ultimate objectives as a result of office politics. It means addressing the vital issues as opposed to those that are merely urgent, having little lasting importance.

Effectiveness is elusive because of the pace and complexity of the environments in which we all work. However, awareness and mastery of a few components along with consistent questioning of one's actions and direction can greatly increase one's effectiveness.

Component 1: Root cause analysis

When faced with a problem, the true root cause may be identified by repeatedly asking "why" questions and applying the test of reasonability to the answers. Let's go back to the bulldozer example and use it as a simple illustrative approach.

"Customers are not buying our new bulldozer."

"Why?"

"Because the competitor's provides a better value."

"Why?"

"They say it has better gas efficiency."

"Why?"

"They don't have to refuel as often."

There are many tools available in attempting to get at the root cause of problems, which will be discussed in later chapters. As has been mentioned, treating the root cause of a problem and not just its symptoms is critical. The more tools in an engineer's tool box, the higher the probability of finding the root cause.

Component 2: The impact of empowerment

Empowerment is one of the most widely misunderstood concepts of our time. Its overplay in the business press and repeated use by management in their speeches and communications has not contributed to a better understanding by the majority of the work force.

At the Saturn division of General Motors, many UAW workers were attracted to its Tennessee plant because of the promise of a different work environment and an opportunity to be empowered. Many came with the illusion that they would be sharing a corner of the general manager's desk and that everyone would have input on all issues, for instance helping determine what the next year's model would look like or other similarly earth-shaking issues. Unfortunately, this is not a true definition of empowerment.

Empowerment more accurately is defining an individual's job as widely as possible and giving each individual the authority and accountability to make the decisions that affect his or her work. It does not necessarily include decisions that their supervisors make. The issue of empowerment can be especially challenging for engineers because at times they work on projects, that if unsuccessful, can put the entire company at risk.

An analogy that has been successfully used to communicate the risks of empowerment describes the people in an organization as the occupants of a ship. They are armed with decision-making "guns," which were formerly held only by the uppermost levels of management. These guns could accidentally shoot holes in the hull of the ship. The challenge is to ensure that everyone is skilled and principled enough not to fire a shot below the water line (Figure 1.1).

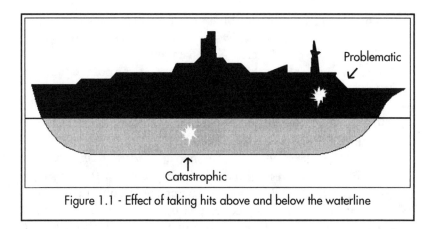

Figure 1.1 - Effect of taking hits above and below the waterline

In our ever-expanding and globally competitive business environment only organizations that have the benefit of discretionary effort from all employees are going to succeed.

Imagine two different employees doing the same job. One does the bare minimum permitted. The other is dynamic and constantly looking for better ways to accomplish more and more tasks. The difference between the two is discretionary effort, the extra exertion made by the second employee. Thriving companies are making a conscious effort to tap this great potential in all their employees.

Empowerment is one of the critical ingredients in eliciting the discretionary energy. However, with increased empowerment, the probability of making mistakes also rises. The higher the number of decision-making guns, the higher the risk of an accidental shot being fired or an intentional shot hitting a vulnerable spot. Successful engineers are the ones who know when a shot they are about to fire has a chance of hitting the ship and what the odds are that the shot may hit below the water line. Does the benefit of making the decision to go ahead with a particular project outweigh the risk of putting the whole company on the line? Who should be involved in the decision-making process? Where is the line at which the engineer can and should make the call, and when should others be involved in the decision-making process? These are questions that can only be answered on a case-by-case basis.

Component 3: Coaching
Coaching is the ability to help people be the best they can be. If we expect people to bring discretionary energy or effort to work, the command-and-control style of management will not work. Command-and-control encourages doing exactly what you are told, nothing more, nothing less. What an effective way to block out creativity!

The command-and-control approach to management is partly responsible for comments from workers such as, "It's not my job," or, "I don't know, I just work here." Not all engineers are in management positions, so why consider this issue? Because, at one point or another in their career, they will work on cross-functional teams, serve as a project engineer or lead a task force. The approach they take can make the difference between completing just the minimum requirements of an assignment or having a huge success.

Meg Wheatley (1992), in her book, *Leadership and the New Science*, brings an interesting historical perspective to the subject of control. She states, "All this time, we have created trouble for ourselves in organizations by confusing control with order. This is no surprise, given that for most of its written history, management has been defined in terms of its control functions. Lenin spoke for many managers when he said: 'Freedom is good, but control is better.' But our quest for control has been as destructive as was his." We believe a coaching and empowering style can make the difference when used instead of command-and-control.

Consultation is an essential part of coaching. When an employee or a co-worker shows up with a problem and asks what should be done, engineers should suppress their training to "solve" problems for the questioner and let their first response be, "How do you think the problem should be solved?" or, "Which do you think is the best solution?" These questions may start the dialogue and be the catalyst for uncovering new solutions.

A Japanese auto executive was asked, "When are you going to worry about competition from the American auto industry?" His response was, "Not until they get all their workers involved in problem solving and innovation." He believed that all the workers in the Japanese auto industry can always beat the few thousand decision makers at the corporate headquarters of their U.S. competitors. There is a synergy involved in the consultation process. As a result of differing points of view and backgrounds, the final product is a much higher quality than it would have been otherwise. One of the great thinkers of our time, Abdu'l-Baha (1982), has said, "The shining spark of truth cometh forth only after the clash of differing opinions." We will address the topic of teamwork in more depth in chapter three.

There is a part of the workforce that is not necessarily excited about the concepts of empowerment and consultation. They would rather be told what to do. When you ask them how a problem should be solved, the response is, "I don't know." The question at this point becomes whether the problem is urgent and requires immediate attention or if it can wait. In the latter case, the best approach is to ask the person who has presented the problem to return when they have developed a couple of alternatives for solving the problem. An individual coached consistently in this manner will soon have alternative solutions developed before they ever present a problem. Businesses are looking for holistic engineers who can raise the problem solving capability of the whole organization.

At coaching seminars offered by the consulting firm Inside/Out, they teach the concept of blocking out interference to improve performance. The concept not only helps with coaching others, it can also be a great self help tool. The basic principle is that everyone has a certain capacity. However, their performance rarely matches their capacity because of interference.

The people who teach this workshop have long been helping sports professionals. A story is told about a basketball coach whose team had won many close basketball games. He was asked how he decides who gets to shoot the ball when only seconds are left in the game. He said he doesn't go to the person who is afraid to lose, even if he is the best shooter on the team because that player will have too much interference thinking about the consequences of missing the shot rather than focusing on making it. For the same reason, the coach also said that he doesn't pick the player who is obsessed with winning. Both players are preoccupied with what will happen *after* the game. The coach's secret to success has been to go with solid players, not necessarily the most talented, who will go out on the court and focus on the clutch shot as if it were any other in the game. In other words, he looks for the player who can block out all of the peripheral interference and concentrate on the task at hand.

There are folks who use this principle in friendly competition to throw their opponents off guard. For example, they ask a friend who is hitting great tee shots whether she is breathing in or out when she strikes the ball. The hope is that the next time she tees off she will think more about breathing than hitting. Or, during a tennis match, they tell their friend how great he is serving and ask how he decides how to plant his feet behind the service line! The whole idea is to introduce interference in the other person's mind and break their concentration.

Sometimes the interference can be self induced. For example, imagine going up a flight of stairs two stairs at a time. Everything will go well until you start thinking of where on the step you are placing your foot, or whether you will trip with the next step. It becomes a self-fulfilling prophecy.

These same concepts apply directly to the business world. Successful engineers are those who can block out interference and increase their performance on a project to the level of their capacity. Companies are looking for individuals who do this for themselves and help their project team members concentrate on the goal they are trying to achieve. Communication skills are not only essential in the success of this process, but also in every facet of the work environment. We have devoted the last chapter of the book to this topic.

Component 4: Vital versus Urgent

In the business world it is very easy to get distracted from the main objective and be drawn to interferences that at the time may seem extremely urgent. The challenge is to be able to distinguish between vital matters and ones that only look like real issues, possibly because of the emotions associated with it at the time or the importance someone else attaches to it.

In engineering school we rarely had to make these decisions. Typically, all of the data in each problem was needed to solve the problem, and all of the assigned problems needed to be solved. In the real world there are many types of data and problems thrown at an engineer and they have to decide which data is irrelevant and what is needed to solve the problem. In addition, the vital issues in many cases are complex, without an easily identifiable solution, thereby increasing the attractiveness of the minor urgent-appearing problems. If urgent problems have relatively easy solutions they become irresistible! Successful engineers should consider the value they add from the perspective of their ultimate customer as well as the department or individual who receives the immediate benefit of their work. This is another case of seeing the forest *and* the trees.

Straightening the deck chairs on the Titanic is not value-added, nor is it the vital problem. However, it can fill some time and one will feel productive in doing it. Uneven rows of chairs may be an easily solved problem, but for what purpose? So, think value-added and ask yourself when addressing a problem: Am I straightening chairs?

Component 5: Ethics

A discussion of effectiveness would not be complete without addressing ethics. In today's world especially, lines that were so well defined in the past have become blurred in the quest to take care of self interests. How many times have you heard people say that if you don't take care of yourself, no one else will? The "Lone Ranger" engineer mentioned earlier in this chapter had apparently taken advice like this way too seriously. We prefer the notion of a holistic engineer, who conveys concern for the whole environment in which one works not just the whole self, understanding that one's success is dependent on the success of others. How can an individual be considered holistic without being truthful, honest, fair, just, and courageous?

A small group of managers were talking during lunch about how untrustworthy a particular individual in their company was. One of the managers, who didn't work as closely with this individual, expressed surprise because he had heard the individual in question was considered to be a religious man. The others said that yes, he was a religious man, but he didn't let that interfere with his business practices. The ethics we profess must be consistent with our actions. Our core values cannot change as we move from the business environment to social and family life.

The business world has been compared to a rat race in which anything goes. However, we should pay heed to actress Lily Tomlin's advice, "Winning a rat race should not be a source of great pride, because it only confirms that you are a RAT!" The solution lies in the refusal to run in a rat race. This is easier said than done, and hence the importance of courage. We must change this perception of the business world through a metamorphosis of our personal lives with an unwavering commitment to the truth. This is not a new idea or a grand discovery. It has been around for ages. Confucius said, "Those who know the truth are not equal to those who love it." Seeking pure truth has been at the heart of all the great endeavors of our race, and honesty a trait common to the most admired figures of history. In a business environment, there is no question that it is often difficult to practice truthfulness and honesty, especially if it could mean the end of one's job or not having enough to make the mortgage payment. However, the choice is ours, whether to be rats in a race or to pursue excellence and act nobly.

As important as fairness and justice are, they are also very subjective by nature. What is fair to one, may not be to another. This is particularly evident in compensation and rewards. Perhaps one worker values their supervisor's expressed approval very highly. Another might see the nature of their assignments as an expression of their value. It seems that for most of us, money speaks and the amount that our salary increases tell us how we are valued by the company. Anyone who has supervised has faced the dilemma of needing to be both consistent in their treatment of all employees and personalized in their treatment of each employee. It can be extremely challenging and frustrating. But, challenging and frustrating or not, fairness must be part of the work environment.

A good start to fairness is provided by the Golden Rule, "Do unto others as you would have them do unto you." Although we can't see the world exactly as others do, if we have the insight to consider their perceptions we will move a long way toward fairness. Through ongoing interaction, we begin to understand what motivates those around us and what they see as positive reinforcement. Through practice, we see what efforts on our part serve as effective motivators. Through providing open communication, we know when a coworker feels they have been treated unfairly. We feel that most perceived injustices in the work place go unnoticed to those who can remedy them. All too often, an employee will present their case to anyone who will listen except their supervisor. In the perfect world we envision, employees would be much more willing to express themselves to their supervisors and supervisors would be much more attentive to their employees.

So what does all of this talk of ethics and fairness have to do with a successful career in engineering? It goes back to the concept of empowerment, which is unalterably based on trust and knowledge. The components that support trust are truthfulness, honesty, fairness, justice, and courage. Without trust, people become fearful and refuse to bring the previously mentioned discretionary effort and energy to the work place.

In her book *Half Life: What We Give Up To Work*, Jeannette Batz (1993) says, "Fear can make people dull or drive them crazy." She quotes Eric Neutzel, M.D., a psychoanalyst, who describes the psychic numbing that takes place when someone can't trust his or her environment, "The person will just do the bare minimum necessary to get by, in a pseudo-compliance without any real emotional investment. As the situation becomes more ambiguous, it becomes more difficult to endure."

If we want people to have an emotional investment in their work and a willingness to bring discretionary effort to the table, we have to treat them with dignity. Timely feedback, whether good or bad news, is part of dignified treatment. If the feedback is negative, it should be communicated such that it doesn't destroy the individual's feeling of self worth.

Engineers may work with many different levels of people in the organization, from the board room to the shop floor. All these people should be afforded the same level of dignity and respect. Engineering's track record in this area is not stellar. It seems engineers have developed a reputation of looking down on people who are not as analytically oriented as they may be. Such a reputation is inconsistent with the concept of dignity and engineers need to do all they can to change it. The level of achievement of the teams you will be involved in is dependent on success in this area. Lack of authenticity and dignity in dealing with fellow workers can be one of the most destructive forces in any environment and especially within teams.

Efficiency

Let's say in your quest to be a holistic engineer you have competency and effectiveness well covered. Efficiency is the remaining ingredient to make your qualifications complete. The age-old saying, "Better late than never," has no place in the electronics industry, for example. In fact, for this industry it is more likely, "Better never than late," and for good reason. Imagine perfecting the high volume manufacturing methods and quality of a PC processor chip just as a competitor introduces the next generation. Or how about manufacturing the best record player when compact discs are introduced to the market. Time is of the essence and successful engineers will recognize the critical nature of efficiency. They will readily realize when they have reached the point of diminishing returns in the analysis of the problem and move forward with a recommendation and follow it with timely implementation.

An interesting observation can be made when measuring the efficiency of white collar and blue collar workers. Typically, the efficiency of blue collar workers will not vary widely, say more than a fifth of the standard time. This would mean that the lowest efficiency would be about 80 percent of standard and the highest 120 percent of the standard. For white collars, the difference is usually much greater, with the most productive employees producing at rates sometimes as much as 100 percent higher than the least productive. It is also the case that the productivity of a blue collar worker is much more apparent, since it is closely measured and monitored. On the other hand, productivity in the white collar group is not readily observable and often not measured. It becomes the responsibility of the worker to ensure that he or she is performing at the optimum level of efficiency.

We have seen engineers who expand a project, which one of their colleagues could complete in five working days, to a month-long effort. We have heard of engineers who fail to bring an assignment to a conclusion unless they are given a specific deadline. These are the engineers who would rather work in a command-and-control environment than an empowering one. The need for these types of engineers is going to diminish and one day disappear because companies they work for will not be able to compete with other companies that employ holistic engineers who understand the concepts of competency, effectiveness, and efficiency. Holistic engineers who possess the core career skills are able to determine the degree of importance of each skill in a particular situation and know which other non-core skills will be required for their success and the prosperity of their organization. The next six chapters will address the core skills.

Chapter 2

Customer Satisfaction: The Ultimate Measure Of Success

A bioengineering firm engaged in the design and development of artificial hip replacements was doing very well and had a very large market share. In order to protect its position, the firm held a meeting once a year at a resort location and invited a number of surgeons using their product to discuss how they could improve the ball and socket artificial hips. At one of these meetings some of the surgeons suggested that a lock system, which would prevent the ball from coming out of the socket, would be a useful improvement. The bioengineering company presented data that in 99.5% of the cases there had been no problem. In the rest of the cases there had usually been some sort of trauma, such as automobile accidents or severe falls, involved which forced the hip ball out of the socket. The surgeons concurred with the data but still asked the company to look into feasibility of this improvement. The engineers further analyzed the data and recalculated the probabilities.

The company decided, in light of such low probability of product failure, it would not undertake a design change. Months later another start-up bioengineering firm introduced an artificial hip with a locking ball joint into the market. Within six months of this introduction, the first company's market-share was reduced by 50 percent and in another six months its parent company, one of the Fortune 500, sold the bioengineering firm because it only wanted to be involved in businesses that were the market leaders.

The moral of this story is that it is critical to listen to customers in order to meet or exceed their expectations all the time. In today's global village there always seems to be some one or some company waiting to catch a supplier who is not satisfying the demands of its customers so they can jump in and take the market away from them, as in the case of the artificial hip manufacturer.

Stories abound about those who listened to customers, provided superior customer service and prospered as a result, as well as those who didn't listen and at the very least suffered great losses and at worst ceased to exist. At workshops and presentations in our company, we used an exercise that was quite revealing to most of the participants. We asked the audience for their opinion as to which of the major retailing companies in the U.S. offers the best service. Invariably, Wal-Mart did very well. (These informal surveys were taken during the 1992-1994 time period.) One of the interesting revelations is that not everyone who votes for Wal-Mart has even been to a Wal-Mart store. But the biggest surprise came when we showed the graph in Figure 2.1.

This exhibit compares the revenues and advertising expenditures of three retailing giants (Fisher 1991). During 1990 Wal-Mart's revenue soared above the other two while it spent the least amount of the three on advertising. The graph demonstrates that reputation for good customer service cannot be achieved through advertising. Satisfied customers are the people who reinforce and perpetuate this reputation mainly through word of mouth and retelling of their experiences.

The U.S. auto industry in the 1970s is often cited as another example of customer dissatisfaction and poor quality leading to loss of market share and profits. Customers wanted high quality cars that

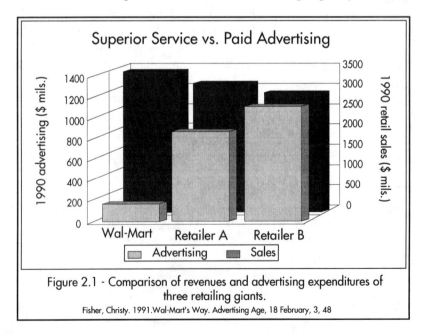

Figure 2.1 - Comparison of revenues and advertising expenditures of three retailing giants.

Fisher, Christy. 1991.Wal-Mart's Way. Advertising Age, 18 February, 3, 48 didn't require frequent repairs, even if the repairs were covered by warranty, as well as cars that offered better fuel economy. The Japanese auto makers responded to this need and benefited as a result.

At a recent workshop on customer satisfaction, one of the participants related the following story of his experience in purchasing a new American-made automobile in the early 1980s. He took the car into the dealership to repair a minor problem that had occurred. They took care of it. A few weeks later he took it back for another couple of adjustments. The service manager told him that next time he should wait until he has ten or more items that require fixing or adjustments before he brings the car in rather than bother them with one problem here and a couple there! A decade later the story almost seems unbelievable in light of the customer orientation adopted by most businesses. However, such occurrences must not have been isolated cases considering the heavy losses sustained by U.S. auto manufacturers. Consumers wanted cars that didn't need adjustments and fixes.

There was a similar situation in the plain paper copying business. Xerox and its army of service technicians would show up quickly to repair broken copiers, but could not compete with copier companies whose copiers were designed to have fewer breakdowns. Quick response to product failure is great, but a fail-safe product is even better.

The Japanese auto makers recognized this fact and delivered what customers wanted. The U.S. car manufacturers have been improving their products in response as well. "*Quality is Job One*" is not just an advertising slogan at Ford, just ask any of their suppliers. The Cadillac division of General Motors won the 1990 Malcolm Baldrige National Quality Award based on the quality of its cars and its commitment to customer satisfaction through providing superior service. Cadillac is now building cars that have their first service maintenance after seven years of use.

Many organizations have learned that having a customer satisfaction goal, as opposed to sales goals, actually results in more sales and creates customer loyalty. Carl Sewell (1990), in his book "*Customers for Life*," claims this strategy helps him attract customers to his Chevrolet, Hyundai, Cadillac, and Lexus dealerships and keeps their business for life. He calculates that this amounts to about $332,000 per customer.

Who is the customer?

In an assembly line, as a product moves from station to station, its worth becomes greater because value is being added at every step along the value chain. As the automobile frame is moved along the assembly line, seats are added and tires are put on. Someone or something adds value by fastening the tire to the vehicle. Along the value chain everyone has a supplier and a customer. Some have suppliers who are external to their company, such as vendors, while some deal with suppliers and customers who all work for the same organization. Finally, there are those who deal with the ultimate customer, the individual paying for a service or product. Often this ultimate customer is the only customer people are willing to recognize and in so doing cause themselves and their companies grief and lost opportunity.

For example, design engineers must consider manufacturing as their customer. Fleet and maintenance personnel should see themselves as suppliers and look at the operating personnel as customers, making sure their needs are satisfied. Everyone has a supplier and a customer and everyone adds value to the final product. The cube in Figure 2.2 is symbolic of the increasing value of the product or service. For an assembly line it is the successive work stations. In a service environment it may be a series of offices or a computer sitting on Jan's desk to a computer in Helen's home.

Along each step in the value chain each individual has to make sure they are adding value for which the ultimate customer is willing to pay and they are doing so in a manner that satisfies the next person in the value chain who is their immediate customer.

Actually, thinking in terms of one ultimate customer can be misleading and overly simplistic. Consider an aircraft manufacturer —

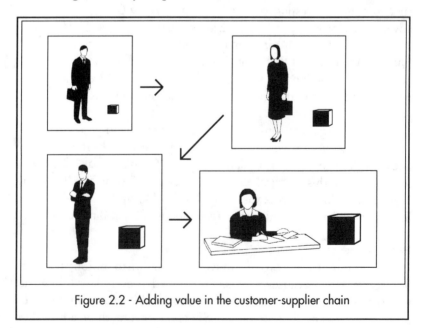

Figure 2.2 - Adding value in the customer-supplier chain

an argument could be made that its ultimate customer is the airline company purchasing the aircraft. A case could also be made that the flying public is the final customer. But, how about some less obvious groups who should be considered customers such as the Federal Aviation Administration (FAA) or the technicians who are going to maintain the aircraft? Anyone having doubts about the latter group qualifying as a valid customer group need only review the history of McDonnell Douglas' DC-10.

On May 25, 1979 American Airlines Flight 191 left the gate at Chicago's O'Hare Field at 2:59 p.m. bound for Los Angeles with a full load of passengers (NTSB 1979) (see also Davidow and Uttal, 1989). At 3:03 p.m., having been cleared for takeoff, the wide-body aircraft started its roll down the runway. At 159 knots, it began to lift off and one second before its wheels left the ground, the left engine ripped off the wing. Each one of the three engines was powerful enough to keep the jet flying by itself, so with the two remaining engines the aircraft continued to climb. However, the crippled left wing was not providing as much lift as the right wing, causing the aircraft to start weaving at an elevation of 370 feet. Thirty-one seconds after takeoff, with the left wing perpendicular to the ground, the airplane plowed into an airfield, killing all passengers, crew, and two bystanders, 273 people in all.

The investigators soon discovered why the engine fell off. The two wing engines are attached under each wing, using a 2,000 pound pylon. The top of the engine bolts onto the bottom of the pylon and the top of the pylon is secured under the wing by means of an L-shaped bracket. One leg of the L is attached to the pylon, while the other leg is inserted into an inverted U-shaped bracket which is then secured using a bolt that passes through the leg of the L and both sides of the U. During the takeoff of flight AA191, the aft L-shaped bracket broke. Investigators discovered a 10-1/2 inch crack on the bracket. Without the support of the aft bracket, the forward bracket also broke allowing the whole engine assembly to rip off the wing. An examination of U.S.-registered DC-10s revealed eight other planes with fractured L-shaped brackets, prompting the head of the FAA to suspend the "type certificate" of all DC-10s on June 6, grounding every DC-10 in the world until further notice. It was estimated the lost revenues cost the airlines up to $7 million dollars a day.

According to the National Transportation Safety Board Aircraft Accident Report, American Airlines had come up with a shortcut in the maintenance procedure for the connection between the L and U-shaped brackets that reduced the required man-hours from approximately 400 to 200 hours. According to McDonnell Douglas' maintenance manual, the connection had to be inspected after every 20,000 hours of flight and lubricated or replaced, if necessary. However, in order to inspect the connection, the engine first had to be disconnected from the pylon and then the pylon removed from the wing providing access to the L and U shaped brackets. American Airline technicians decided they could save considerable time by not detaching the engine from the pylon. They supported the engine using a forklift, detached the pylon from the wing and lowered the engine and pylon together. After the inspection and necessary maintenance, the vertical leg of the L was raised into the U shaped bracket, which had a clearance of less than an inch. American Airlines had developed this procedure with the knowledge of McDonnell Douglas and had performed it over forty times without any apparent problems. The FAA investigators determined this procedure was too imprecise. Damages to the L-shaped bracket caused the crack, resulting in the crash, as well as cracks in other DC-10s maintained using this procedure.

Had the engineers at McDonnell Douglas thought about the maintenance of the aircraft from their customers' viewpoint and the impact of 400 man-hour procedures, they may have modified the power plant assembly. The problems with the serviceability of the DC-10 may have been the major contributor to McDonnell Douglas losing its number two position in commercial aviation.

Aircraft manufacturers are not the only ones who have to think of building service into the design. A few years back a story was told about a man who wanted to change the spark plugs on his car. He replaced seven spark plugs and could not find number eight, which he knew had to be there somewhere. Like many of us, after all else failed, he went to the maintenance manual which informed him the number eight spark plug is easily accessible once the engine is removed from its mounts and elevated by two feet!

The point is that companies are looking for engineers who design products with an eye on all customer groups—from users of the product, to those who service and maintain it, as well as the regulators who may have an impact on the success of the product and the company in the marketplace. The key is to develop an exhaustive list of all potential customers along the value chain, whether internal or external to the organization and to make sure that each one is satisfied.

Determining customer satisfaction

A teenage boy was sitting at the lunch counter of a local cafe when he walked over to the public phone by the cash register to make a call. The owner of the cafe could not help overhearing the conversation. It seemed the young man was attempting to get a new customer for his lawn mowing business. He told the party on the other end of the line the services he offered and was apparently told they already have someone who is mowing their lawn. However, he did not give up. He mentioned how reliable he was and he provided service with a smile. He mentioned as many advantages as he could but was unable to persuade the homeowner to switch lawn mowing services. The cafe owner was impressed with the teenager's perseverance and selling skills and consoled him by saying he couldn't help overhearing and he was sorry the boy couldn't get the account. However he was confident there would be many others who would use his services. The boy responded, "I already have this account, I was just calling to make sure they were satisfied with my work!"

There are many different ways of determining customer satisfaction, including surveys, complaint and product return analysis, competitive position analysis, and actively and aggressively pursuing understanding of customer expectations. To be useful, however, they all require a commitment to the belief that individuals as well as organizations need meaningful information about how well customers are being satisfied in order to be successful in the long run. The only way to gain this meaningful information is by asking the customer. Satisfaction is an elusive and dynamic commodity and hence the critical importance of asking. Customers who were satisfied with Xerox's prompt and knowledgeable service technicians one day, became dissatisfied when they became aware of copiers that didn't require as much service. An individual customer's satisfaction with a product or service is determined not only by the product or service itself but also with the expectations created in the marketplace by competitors. All airline companies in the U.S. had to offer a frequent flyer program soon after American Airlines introduced it's program in order to satisfy their customers and retain their business.

While direct contact with the customer and inquiry about his or her level of satisfaction is most effective, there are other measures that can also be useful. The challenge is to make sure the measures are what the customer uses to measure the product or service. Some companies tend to use measures they know the customers will rate well while ignoring or not even inquiring about the areas in which they perceive they are not doing as well (even though the latter is a measure used by the customer to compare the provider with its competition). This shortsighted approach creates a false sense of security, which at best is detrimental and at worst fatal in the long run.

Customers, whether internal or external, care about measures such as wait time, delivery time, and response time. Other reliable customer focused measures are the number of repeat customers, as well as new customers referred by existing customers, results of customer satisfaction surveys, and number of complaints.

Some may argue that these measures are applicable to sales and service departments and are either not available or nonexistent for engineers. Successful individuals collect the above mentioned type of information even if their department or company does not. They are cognizant of how long they take to respond to a phone call, voice mail, or electronic mail. They are the engineers who try to minimize the wait time of their customers—whether it is the sales department looking for a prototype or the manufacturing department asking for a process modification or the budget department waiting for an estimate on the cost of a complex proposal. They are the ones, who after completing a project, follow-up to see if the customers needs were met through a survey or a simple phone call. They inquire how to better meet their customers expectations the next time they do business. They keep a mental note, if not an actual count, of how many people have asked for their help with this or that project based on the recommendations of friends in the company and finally they keep track of repeat customers, as well as the number of complaints.

We all have customers and using customer-oriented measures in combination with direct feedback will provide a powerful competitive edge. It will serve as the basis for the development of a list of actions necessary to maintain and increase customer satisfaction. Successful individuals and their companies analyze and synthesize all available information to anticipate the future expectations or unspoken needs of their customers.

We mentioned customer complaints as a valid customer satisfaction measure. It is important to point out that the number of complaints by itself is an extremely poor measure of customer satisfaction because most unhappy customers will not complain. Think about the last ten times you were unhappy with a service or product. How often during those ten experiences did you complain to the provider? Most people don't, however they do manage to tell a number of their friends and associates about the bad experience. According to a study by the White House Office of Consumer Affairs, 96 percent of unsatisfied customers will not complain but they will on the average tell eleven other people and 91 percent will not do business again with the provider of poor service. The number of complaints is a good measure of customer satisfaction if it is NOT the only measure being used. If complaints are the only yardstick of customer satisfaction in your organization, help the organization realize the dangers involved and if you don't succeed in changing the approach, you may want to think about finding another job!

A personal commitment

Customer satisfaction requires a personal commitment. Customers can see through facades and empty slogans. They can tell when "talk" is not "walked." What is required is a new approach to the way we think about work. Engineers, by training are supposed to use science and technology for the betterment of humanity but it seems too many of us are not thinking about that when we go to work everyday. If we see work only as a way to make a living, customer satisfaction is just another burden imposed on us. However, if we consider work as a service to mankind, customer satisfaction becomes an integral part, if not the whole objective. Considering work as service leads one to look for ways to add more value and adding value not only contributes to customer satisfaction but also to a feeling of having accomplished something worthwhile as opposed to just being busy. Career advancement is also another by-product of being focused on adding value for obvious reasons.

This alternative approach to work is even more critical for success in organizations that have made customer satisfaction a top priority— companies whose structure has been transformed from functional silos or chimneys to networks formed to better serve customer needs. Classical organizations are clusters of functional silos, the research and development silo, the finance silo, the manufacturing silo, the sales and marketing silo, and so on. The networks in the transformed companies perform quality control through prevention rather than error detection. Their priorities are finished goods at the lowest possible cost and the highest quality and value. Their response to failure is continuous improvement as opposed to fixing blame. Their approach is collaborative not adversarial. They provide leadership for their teams instead of directing them and their expectations from employees differs from a traditional organization. They look for people who have multiple skills, not just specialists; people who not only work alone but also contribute and add value to a team; people who don't have to be told what to do and how to do it but rather those who know what is needed and can develop a plan to achieve the required results.

In these transformed companies, a personal commitment to customer satisfaction translates into providing the customer with what has been agreed upon—every time. A 100 percent customer satisfaction goal may seem unrealistic at first glance, especially to engineers who have been trained in Acceptable Quality Levels (AQL). Usually the

AQLs allowed for a certain percentage of defect, albeit a small percentage, because it was considered uneconomical to inspect for 100 percent conformance. However, to design for total conformance to specifications is possible as well as economically feasible. A large computer company ordered some parts from a Japanese manufacturer on a trial basis and in the specifications included an AQL of 0.03% or three defective products per 10,000. The order arrived in two packages, one large and one small. Attached to the invoice was the following note:

"We Japanese have hard time understanding North American business practices. But the three defective parts per 10,000 have been included and are wrapped separately. Hope this pleases."

Meeting customer expectations 100 percent of the time can pose many varied challenges. An engineering director from an electronics firm wanted his engineers to deliver the product to the customer once the design met all of the customer specifications. He complained of situations where the engineers became so enthralled with the technology to the point of adding features and extra bells and whistles not asked for by the customer. These extra features not only added to the cost of the product but also cost the customer time. In the electronic industry where the motto is "Better Never Than Late," time is a precious commodity and every incremental hour spent in time-to-market can make a sizable difference in the success of a product. The engineering director wanted his staff to deliver the design to the customer as soon as it met all the criteria and then consult with the customer about the possible additional features and enhancements. They could then jointly determine if the enhancements were worth the extra time and money.

In order to have fruitful consultations, the customer needs to have a good understanding of the processes of the provider. Engineers need to explain to their internal or external customers how they do things and explain the process in the simplest possible terms. There was a time when you could tell a customer we put a few items in this black box and the answer to the problem jumped out after the proper gestation period. Today customers are becoming more sophisticated, knowledgeable, and less willing to accept the "black box theory." They are likely to choose the service of engineers or engineering firms that can demonstrate they have the systems in place and how the proper use of these systems produces results required by the customer.

A fine balance

Most customers are reasonable and fair-minded. However, there are those few who seem never to be satisfied, constantly asking for more. If you give them an inch they want a mile. Our advice with this class of customers is to give the inch with a smile and then negotiate beyond that. "Giving an inch" with a smile has a much greater impact on customer satisfaction then giving it begrudgingly, even though the same amount of effort is expended in both cases. There is a fine balance between satisfying one particular customer and being taken advantage of. Each individual needs to determine this balance and when in doubt, err on the side of being taken advantage of. It is not prudent to let this very small class of customers stand in the way of superior service. If one agrees to be taken advantage of in a particular encounter with a customer, doing so with the right attitude and a smile can turn a potentially negative experience into an extraordinary customer service event.

Another area requiring a fine balance is follow-up with customers. Follow-up is a key to success in customer satisfaction and should be performed such that it doesn't become a nuisance to the customer. Long surveys and telephone calls at inappropriate times may actually do more damage to customer satisfaction than the benefit they are supposed to provide. Successful engineers make it easy for their customers to provide feedback. They know when to use a formal survey instrument that contains a few key questions and when to just ask the recipient of a report if it met his or her needs during a break at the company cafeteria or at the drinking fountain. Others design their own surveys focused on action not just satisfaction. Their survey has only one question:

"How can I better meet your needs the next time we work on a project?"

Chapter 3

The Value Of Teamwork: It's A Symphony, Not A Solo

In a symphony orchestra that is performing at its peak, it is sometimes hard to distinguish between the various instruments, but what you hear is magical! And when you do hear the flute, violin, or piano, it is by the design of the composer and it complements the performance of the symphony as a whole. Engineers can learn a great deal from musicians in this regard. The days of heroes who parachute in to fight fires and solve problems may be numbered, if not over. There will still be fire fighters but the real heroes will be those who cooperatively design systems that eliminate problems and prevent fires from starting.

Survival in today's economy necessitates concurrent engineering to reduce "time to market." It demands reduction in cycle times to produce high quality products in shorter time periods. Typically, this can only be done in a cooperative team environment.

One of the authors was a martial arts student for quite a few years and one of the first lessons taught by the Sensi (master) was about a martial arts master who was so good that he could take on ten attackers at one time and prevail over all of them. One day eleven opponents challenged him and beat the living daylights out of him! The lesson the Sensi wanted the students to learn was not to go around picking fights with their newly learned skills. The moral of the story from an engineering point of view is that no matter how good an engineer may be, if he or she is going to go the "Lone Ranger" route, success will be elusive if not unattainable.

Let's say you have identified a need in the marketplace and you think you can design a widget to fill the need and create a successful organization as the result. You proceed with the design of the widget and spend some time refining the plans. Once you are happy with the quality of design, you start looking for a manufacturing facility. One plant manager reviews the design, tells you how much it will cost you to manufacture the widget and recommends a number of changes to your design to reduce the manufacturing costs and improve the quality of your product. The changes seem reasonable but they are going to take some time and you know that others must also be thinking about producing such a product, so you decide to look at additional facilities to see if one can be found that can match the price but be able to proceed with the current design. The search takes you to a number of different plants and visits with plant managers and manufacturing engineers. Finally, you find one who cannot quite match the price of the first plant, but is willing to proceed with the current plans. The night before you are supposed to sign the contract and give the second mortgage on your house to the manufacturer, a call comes in from a friend with whom you have discussed your idea to let you know another company has introduced the widget in the marketplace. How much do you like your house? What is your confidence level in the design and quality of your widget? Can you catch up with a competitor who beat you to the market? How many more of them are there? What resources are at their disposal?

Companies have gone to concurrent engineering and the team approach not because it is easier, but out of necessity. The sequential approach to product development was probably simpler but took a long time. Today organizations view time as a scarce resource. In addition, the team approach results in a synergy nonexistent in the "over the wall" system of product development. As soon as a need appears in the market, companies across the U.S., Japan, Singapore, and Europe put together cross-functional teams to meet that need. Customers demand a rapid response to their requirements with a quality product or service. In today's global village, there is someone out there who can deliver just that. They do it by bringing people together from various backgrounds and perspectives who are customer-focused and realize the importance of delivering what the customers want as well as anticipating their changing needs. Companies need people who are comfortable with disagreement, who build consensus, who resolve conflicts in a timely manner, and who are doggedly focused on their goal so that their performance will closely match their capacity.

When to organize a team

Usually engineers are assigned to teams and do not have to worry about this question. However, there are those times when a specific assignment is given to an engineer without any mention of a team or a team approach. Complexity of the assignment, the deadlines associated with it, the number of processes and people involved in it, as well as affected by it, are all variables that should be considered before making a decision to bring a team together. The higher the number of people and processes involved in an assignment, the higher the need to organize a team. We have seen numerous projects and recommendations derailed because the engineer in charge did not get the buy-in of a group of affected people and as a consequence their respective vice-president shot the project down because it was felt the recommendations were not workable. People do not necessarily resist change, they resist being changed without having a say.

An acquaintance of ours was given an assignment by the president of his company to study the signature approval policy of the organization and develop recommendations to improve it. The goal was an expenditure approval policy befitting an empowered and accountable workforce. The scientific approach suggested gathering data about what was happening by taking a sample of past invoices from the accounts payable, office supply, purchasing departments, etc. The data was analyzed, opportunities for improvement identified, and recommendations for action submitted for approval.

Our colleague took this general approach, but in addition, involved a number of people from all major functional areas of the company. He knew that a signature policy would impact the whole company and it was very near and dear to a lot of employees. He decided to involve them right from the start. He asked each department to develop a list of all items that require approval and a description of the current level of signature authority for each item. In addition, for each item on the list the respective manager was asked if he or she would like to change the level of authority and why the change was necessary. The final request was to suggest that an individual from each department serve on the team. Recall that the assignment was given to one person, but he realized that in order to be successful, he needed input from a fairly large group and took the initiative to pull a team together to work on it. The advantage of this approach over the sampling of old records was that he not only learned what people were currently doing, but also how they thought things ought to be done in order to better serve their customers.

At the first team meeting it became clear that the recommendations for change were not consistent. Each representative got the opportunity to explain the reasoning behind his or her recommendation. The ensuing discussion on the advantages and disadvantages of consistency as opposed to flexibility proved very enlightening to everyone. The group finally came up with a set of recommendations that everyone could agree with. When the final draft was presented to the president, it had the support of the executive staff and their departments. Although no one liked every aspect of the proposal, they understood why it had to be that way and why they chose to support it. Contrast the chances of success for this proposal over one developed in isolation, based on sound reasoning, approved by the president, and communicated as a list of "Thou shalts."

The key consideration in the above example is the probability of success, not the amount of work or time involved. We readily concede that in most cases it is harder to take the team approach and it will probably take longer to find a suitable solution. Achieving consensus is hard work and requires skilled and principled people. However, once consensus is achieved, the probability of success is very high.

At the Saturn Division of General Motors, management is making every effort to ensure people receive enough training to be skilled in the specific technical aspects of their jobs, as well as how to work in teams and make decisions. In a recent visit to the Saturn plant in Tennessee, we asked the co-team leaders of the chair assembly group how the group decides which direction to pursue when faced with problems. They said they implement solutions that have the support of seventy percent of the people in the work group, but they have an agreement among themselves that they will all be 100 percent committed to making the agreed-upon solution work. Their success confirms again that skilled and principled people are indispensable assets.

Who should be on the team

Sometimes team members are like family, you don't get to choose them. In these situations, assess the strengths and weaknesses of each member and determine if there are any needed skills that are missing. If possible, add one or more members who bring those skills. If not, be aware of this vulnerability. Constantly check and recheck this area and compensate for it every step of the way.

Teams should be brought together when faced with changing or improving a process or implementing a project. The ideal team would include a participating representative from all affected areas. However, if this results in a large team, the sheer size will most probably be a handicap for the group. The preferred number of members on a team is between five and nine. Odd numbers seem to work better because it is impossible to have a tie vote. Of course, this is based on the assumption that no member can abstain from voting. They are either supportive of an idea or against it. But what if there are twenty areas that are affected and they all should be involved? One solution is to form sub-teams whose leaders serve as the liaison to the main group of people charged with the project or assignment.

Another consideration is to include members with a 360-degree view of the world as opposed to a gun barrel view, and who are willing to express that view. The worst thing that can happen is to end up with a bunch of "yes-people" on a team.

Figure 3.1 is a list of qualifications for team members developed expressly to combat the problem of compliancy (Katzenbach and

---

Technical skills:
Knowledge and experience relevant to the project

People skills:
Ability to work effectively in a particular group

Commitment:
Interest in the project and ability to carry through

Figure 3.1: Considerations For Selecting Team Members

Smith, 1993). The "yes-people" are from a tribe that believes the way to get ahead is by feigning subservience to the person in charge. They are trained to stroke the ego of the person with authority at every possible opportunity, regardless of the merit or consequences of suggested ideas or actions. One of their other talents is the ability to detect the direction of political winds and change their opinion accordingly with rapid speed. Unfortunately, in many organizations, these folks have been very successful. They have also been the downfall of organizations and even communities. It seems the human race keeps falling into the same ego trap over and over again because people in positions of power enjoy having individuals around them who are agreeable.

There are a few lessons to be learned here for all engineers. First, be aware of the existence of this tribe and look for them on your team. Sometimes they are hard to identify and they always disguise their affiliation with the "yes-people" tribe. Second, guard against becoming a leader who is dependent on the members of this tribe. If you feel you are the only one with good ideas and solutions, and your team members agree, you are surrounded by "yes-people." As a result, your function and company will have competitive blind spots, and lack the 360-degree viewpoint. Third, the intent is not just to put a team together and then push your own solution or agenda like a bulldozer. The idea is to achieve the synergy discussed earlier: to have the group come up with solutions that no one individual could have done on their own. The way to achieve this synergy is through the art of consultation, a discussion of which will follow later.

Dream team

Once in a while one has the good fortune of being on a "dream" team. Sometimes you get to pick your teammates for this team, and on rare occasions, you just get lucky and end up on a team with a group of holistic professionals—people who are competent, effective, and efficient; people who feel and are indeed empowered, skilled, and principled; authentic individuals who understand and believe in the concept of treating people with dignity. When you have a team like this, the sky is the limit and the possibilities are infinite. We have had the pleasure of being on such a team and experiencing the magic it can produce. What follows are the recollections of a member of one such team.

"I was assigned to a cross-functional team that was to develop a computerized merit review system for the company. A larger group had already designed the basic content of the review process, and we were to create the associated system as well as a procedure to equitably distribute the budgeted salary increase to employees based on the performance reviews. This system was to be automated and allow the supervisor to complete reviews on-line and print the necessary reports.

"The group included representatives from finance, operations, human resources, data processing, information services, and engineering. At the time, I knew we had a pretty good team, and in retrospect, it was indeed a dream team. We all shared a clear vision about our objective. We wanted a system that was fair and just to all employees and gave supervisors flexibility in rewarding performance with pay. There were no hidden agendas on anyone's part. When the team was formed, I did not know the systems analyst representing information services, who would lead a team of programmers in coding the computer system. This was one of the times we just got lucky and ended up with a holistic systems analyst on our team. (You see, it is not just holistic engineers that are in demand. Organizations are seeking holistic employees in all fields.)

"We started down the general path recommended by the compensation task force and along the way we learned about each other's strengths, talents, and philosophies. When dealing with salaries, most people have very strong beliefs and are quite opinionated. These strong opinions resulted in many heated discussions. After a few team meetings, people with offices next to the conference room would ask what the shouting match was all about. The team had developed such a level of trust that it was acceptable to be audibly passionate about a point and still retain respect and credibility. Disagreements happened often, but we worked on them until they were resolved.

"The most memorable disagreement took place a number of weeks after the project had begun. We realized the methodology on one issue suggested by the parent task force would cause serious problems if implemented. We put our case together and invited the project sponsor to a presentation. We succeeded in getting his blessing in changing direction. He later told us he had come in ready to tell us the change was beyond the scope of our assignment and we should continue on the original path, but realized the group had uncovered a real issue and was grateful to have dodged that particular bullet.

"When I think about the team members and use the holistic measuring stick [outlined in chapter one], I can see clearly why they were a dream team. They searched for root causes and were able to distinguish them from the symptoms. They understood empowerment and recognized when they were about to fire a shot below the water line, as in the case just described. They knew how to coach each other in blocking out interference so each individual's performance could come as close as possible to their capacity, thereby maximizing the team's performance. They stayed focused on the vital issues and didn't get distracted with urgent-appearing matters. But above all, they conducted themselves with the highest standards of ethics, which I believe was the catalyst for the extraordinary level of trust among team members.

"One of the characteristics of their ethical conduct was treating everyone with dignity. I recall one instance when my colleague from operations was almost yelling at me from across the table. The image of him with a flushed face and wide-open eyes staring at me is etched in my memory. He and I were engaged in one of our debates and he wanted to make sure I understood his feelings about the subject and why he felt I was missing the point. At no time, even during these heated debates did I feel that my dignity as an individual was attacked or harmed. It was the clash of differing opinions that ignited the spark of truth for our team."

Forming a dream team

Some organizations attempt to pull together team members who complement each other's styles and ensure a 360-degree view of the problem or project at hand. They sometimes use a personality profile instrument to gain a better knowledge of each individual's pattern of behavior. One common and generally well-accepted instrument is the Myers-Briggs Type Indicator (MBTI). Based on a person's answers to 126 questions, the MBTI classifies the individual into one of sixteen types. Each type describes the way the person prefers to handle situations they encounter, decisions they make, and how they deal with life in general.

Four activities are considered: "Energizing," "Attending," "Deciding," and "Living."

Energizing describes the preferred sources for the respondent's energy and information. This dimension measures extroversion/introversion, revealing whether the person tends to build associations or define territories, prefers talking to writing, is "good" with people or prefers to work alone, etc.

Attending refers to the person's perception of reality and what they are most likely to consider when gathering information. It reports on a scale of sensing/intuition, and separates those who see details from those who prefer to see the big picture, prefer to live in the present or future, see what is or what could be, and so on.

The third activity is *Deciding* (thinking/feeling scale), and it characterizes the person's preferred models for decision-making. This indicates a preference for justice or mercy, laws or circumstances, following their heads or following their hearts.

The final descriptor is *Living*, and on a scale of judging/perceiving, it reports the person's preferences for dealing with other people and change, whether they express thoughts and feelings or senses and intuition, whether they are planners or are flexible, and whether they prefer to have loose ends tied up or are willing to leave things open-ended.

As with other personality indicators, there is no right or wrong or better or worse among different personality types. One type may be very good in a leadership role but cannot function as well when they are a team member. In fact, the MBTI suggests that the need for certain types is situational and that teams will need a combination of several of the sixteen types.

Imagine a team where all of the members had a disposition to see the big picture rather than the details. They would probably be very successful in making general plans but not as effective in implementing a course of action.

The challenge for the person developing a team is to include members that will give the team a composite personality capable of meeting its goal.

The art of consultation

Observing the art of consultation at its perfection is usually a rare event because too many factors can interfere and keep the process from achieving its capacity. However, many try to consult on issues through staff meetings, team meetings, and other meetings. These efforts are beneficial even if they do not produce all of the value that is available. Some consultation is better than none at all. Everyone is looking for the elusive synergy the consultation process offers and the resulting generation of ideas, which otherwise would go undiscovered. Once again, this is not an earth-shattering discovery. It has been around for ages and it has led to many scientific discoveries. Peter Senge (1990) refers to it as the discipline of team learning in his book, *The Fifth Discipline.* He quotes physicists Werner Heisenberg and David Bohm, who each describe the same condition. Heisenberg recollects his associations with Pauli, Einstein, Bohr, and other pioneers who reshaped our understanding of nature. Heisenberg said, "Science is rooted in conversations. The cooperation of different people may culminate in scientific results of the utmost importance." Many of the separate achievements for which these scientists are recognized were the result of this ongoing collaboration. Bohm refers to the practice of dialogue that was common among the ancient Greeks and Native Americans. These dialogues take on a life of their own, and the creative synergy provides the participants with previously unimaginable ideas.

Consultation needs to become part of our systematic approach to problem solving as well as idea generation if we want our organizations to remain competitive on a global scale. However, this is much easier said than done because the promise of synergy in the consultation process is conditional upon a few key factors.

Letting go of your ego

As a team member one of the hardest things to learn and master in the art of consultation is the ability to let go of your ego. It seems most of us feel obliged to support and defend an idea we have put forward, right to the bitter end. Often we fight for ideas that, given a chance to develop and carefully consider, we don't really believe. The work of the team can be undermined if team members are guided by self-interest. There has to be a purity of motive, considering the real issues that face the team from the perspective of all involved. This problem may be avoided by remaining open-minded in disagreement, and exploring the reasoning of a colleague with an opposing point. Once you understand your colleague's opinion, it's fair to develop or present your own.

The effort to remain open-minded is accomplished by detaching yourself from your idea once it has been put forward. That is to say, an idea put forward should be considered the group's idea, open to a free and critical examination by all. Even though this approach may seem odd at first, with practice, group members can put it to good use. The advantage of detachment from ideas is that it helps the team members keep their egos out of the process and not subject themselves to emotional bruising.

It must be strongly stressed that detachment does not mean a lack of support of an idea that has been presented. On the contrary, ideas should be supported diligently with the facts and data available and argued with the most systematic reasoning possible. But this must always be done with an open mind. There is a fine balance between advocating a position and being open-minded enough to recognize the shortcomings of the position and encouraging adjustments to the point being made.

The opposing side of detachment from ideas is defensive routines. When conflicts arise and the people involved are unwilling to openly discuss the issues and alternatives, they resort to these defensive routines. Defensive routines are habits that everyone develops to protect themselves from ridicule and embarrassment, but they also inhibit one's ability to learn from others. Chris Argyris (1985) argues in his book, *Strategy Change and Defensive Routines,* "We are programmed to create defensive routines and cover them up with further defensive routines."

Too often in group work, members are reluctant to express themselves because they are afraid that they will be ridiculed. (Depending on one's self confidence, ridicule could be any response less than adoration.) Senge (1990), in presenting team learning, says that defensive routines are a deeply-rooted part of our psyches, developed to protect us from being exposed or hurt. Unfortunately, this protection comes at the cost of our being able to learn and reason.

As difficult as it is to confront demanding issues, we have not experienced anything that is intellectually more exhilarating than participating in the discussion of honest and intense disagreement with a trusted coworker. Trust is important since it allows us to abandon our defensive routines.

In trying to avoid defensive routines and resolve conflicts, Senge (1990) offers the following suggestions.

When advocating your view:

- Encourage others to explore your view;
- Encourage others to provide different views; and
- Actively inquire into others' views.

When inquiring into others' views:
- Vocally recognize assumptions you make about their views;
- Share the data upon which your assumptions are made; and
- Pursue discussion only if you are actually interested in the other person's response (not motivated by politeness or competitiveness).

When you arrive at an impasse:
- Ask what data might change their views; and
- Ask if some experiment might add light to the discussion.

When you or another is hesitant to express views:
- Vocally recognize the hesitancy and seek to understand it; and
- Work with others to break down these barriers.

Tools for getting the most out of teams

Brainstorming has become a commonly used method to solicit ideas quickly from team members. To begin a session, clearly define what is driving the discussion. The driver could be a problem that needs resolving or a question for which the best answer is sought. Once this driving issue has been defined, write it where everyone in the group can see it. This will help keep the suggestions focused on the issue. For example, the issue might be, "How can we accommodate the parking needs of new employees?" or "Why have sales of the new product line been better than expected?"

Once the issue has been clearly stated, give team members a few minutes to prepare their thoughts, encouraging each to come up with several ideas. Then begin gathering ideas. The leader must make it perfectly clear that this is to be done without discussion, including calls for elaboration, amendments to suggestions, and especially without criticism. It is often useful to appoint one member of the team to monitor the process for these infractions. The intent of brainstorming is to take advantage of the group's synergy, and let them think both collectively and individually. Ideas are developed and built on by members of the team as one team member presents them. Every idea is written on a flip chart, using the words of the person who presents it. Members can either present ideas in a "free-for-all," or ideas can be collected one at a time, with each member in turn having the option of giving an idea or passing. The latter method is called "Round Robin," and works especially well to give response time to reluctant members in groups where a few of their peers tend to dominate (This is the case in most groups).

Multivoting provides a means to condense a large list of ideas down to a smaller list of the most important. The process begins with a numbered list of items to be considered. The list may have been developed with brainstorming. Items on the list may be combined as long as the group agrees they are identical in content. Members of the group vote silently, either by show of hands or by ballot, on the items they feel are most important. Each member is allowed to vote for about one-third of the total items on the list. Once votes have been tallied, items on the list that received few votes are eliminated and the process is repeated until only a small, manageable number of items remain.

Nominal Group Technique (NGT) is a structured process by which a team reaches a practical answer to a complex or controversial question by a fairly direct route. Because of its structure, it is particularly good in situations where team members are not familiar with each other or comfortable with the topic at hand. Even so, it works well in most team situations, and as the group becomes familiar with the process, it will become almost second nature and will certainly enhance the effectiveness and efficiency of their efforts.

NGT begins with a brainstorming session as described above. It is important to clearly define the purpose of the discussion and have it posted where everyone can see. Team members begin by developing ideas using the silent generation method and gathering them using round robin, writing them down on a flip chart.

Once the ideas have all been written on charts, team members ask contributors for clarification where needed. Where more than one item is *clearly* identical, combine them. This can be done by cross-referencing the items on the charts so that everyone knows which items belong together. In any case, changes can only be made with the consent of the contributor(s).

Members then vote on the items they feel have greatest impact on the defined purpose of the session. The number of votes for each member should be equal to about one-fourth of the total number of items on the list (The list should typically be no more than 50 items). For instance, if there are 20 items, each member may select the five they feel are most important. To vote, each member receives numbered (one through five) 3x5 cards or Post-it® notes. (Again, let's say they each get five votes, so each gets five cards.) To keep things straight, it is best if these numbers are circled. Then, on the card with the highest circled number, members write the number of the item they feel is most likely to resolve the problem being considered. It is helpful if they also write a short description of the item. Next, members select the second most important item, write its number on the card with the second-highest number, and so on.

The scores for each item are tallied, and the items are then ranked according to their score. In cases where two items tie in score, the number of people who voted for the item is used as a tie breaker. For example, say items 12 and 26 both received votes totaling sixteen, and four people cast votes for item 12 and five members voted for item 26. Item 26 would then receive the higher ranking.

An important step in NGT is to review and discuss the results. Members should be comfortable with the rankings. We experienced one case in which the item with the highest ranking turned out to be everyone's second choice, and the members in general felt this item was important, but alone would not have a great impact on the problem. After considerable discussion, another vote was held from which a new ranking emerged.

Case in point: The hiring process

The team approach may be used in many different situations and in fact, the tools are just as useful in nontraditional applications. In chapter one, we discussed finding holistic engineers and the interviewing process we use in our department. This process begins with a round of NGT to help identify which of the several applicants should be interviewed. In the past couple of years we have had four openings in our engineering department and every time we have been fortunate enough to have at least thirty candidates apply for each position. In one case, over fifty individuals applied for an opening.

We begin the process by collecting their resumes and applications and routing them to everyone in the department. A three- to four-member selection team is then chosen from the group. After deciding how many applicants each member may vote for, typically about five, they review the applicants and make their top choices. The data is compiled and a summary sheet is prepared. Let's assume that we had agreed each list should include five names. Each selection team member's top selection is given five points, their second choice four, etc. Using the resumes and this ranking, we develop a short list of the best candidates (see Figure 3.2). This is the silent idea generation step of the NGT process.

After team members have reviewed the summary sheet, we hold a meeting in which they have the opportunity to ask questions about each other's assumptions and perceptions. We discuss why an applicant may be on all lists except one. We consult whether all the candidates

Candidate Summary Sheet

Lauren	Danielle	Lawrence	Renae	Bijan
1.John G.	1.Navid A.	1.Ben T.	1.Anna C.	1.Jessica R.
2.Anna C.	2.Milo B.	2.John G.	2.Nikka E.	2.Ben T.
3.Ben T.	3.David T.	3.Farbod H.	3.Navid A.	3.Mona K.
4.Jessica R.	4.John G.	4.Milo B.	4.David S.	4.Anisa S.
5.David S.	5.Farbod H.	5.Anna C.	5.David T.	5.Gloria T.

Top Ten Candidates

Rank	Name	Points
1.	Ben T.	12
2.	John G.	11
3.	Anna C.	10
4.	Navid A.	8
5.	Jessica R.	7
6.	Milo B.	6
7-9.	David T.	4
	Farbod H.	
	Nikka E.	
10.	Mona K.	3

Figure 3.3 - Candidate Summary Sheet.

who should be on the short list actually made it following the voting. Team members get the opportunity to speak for candidates who didn't make the short list. We look at the spread of points to see where there is a logical grouping based on the points given to the applicants. We don't rely solely on the points. We review the results and see if our common sense tells us whether an applicant was excluded inadvertently. Tools such as NGT are very helpful, but cannot replace sound judgment.

The applicants in the top group, three to five in number, are invited in for some rather intense and revealing interviews. After all applicants in the group have been interviewed by all members of the team, the selection process is repeated and a job offer extended.

Admittedly, our sample size in this experience is not large, but the process has been 100 percent successful. In all four cases, we found talented, holistic engineers who have been great additions to our team. In three out of the four selections the teams were unanimous in their vote. The lack of consensus in the other case could have been due to our lack of skills in the art of consultation or the inability of some team members to detach themselves from their egos so they could consider the department's needs over their own. In any case, we were able to better understand the process in subsequent efforts and have enjoyed a very positive experience in each case.

In chapter one, we talked about the critical role that trust plays in the proper functioning of a group of people who work together. "Walking the Talk" (a consistency between what one claims to believe and what one does) is an important contributor to the establishment of trust. In the second hiring process for the department we had an interesting example of this concept. The director of the department had invited an engineer he knew from a professional association to submit an application for the opening. This individual made it to the interview round and was one of the finalists. After the usual round of consultation, all but the director had abandoned support of this candidate. One member of the team, who had just one year of tenure in the department, later said that he was certain that this candidate would be the one offered the job despite the team's selection. He was surprised to find that by the end of the consultation the director fully supported and went with the team's choice instead.

Chapter 4

Managing Technology: A Means, Not An End

From the day a Neanderthal first used a rock to drive stakes into the ground to yesterday's revision of the microprocessor, human kind's marriage to technology has been both rocky and irrepressible. Depending on the circumstances, we each may fear, flee, decry, and disdain technology, or we may trust, pursue, applaud, and embrace it. Of course, much of the time we don't even notice it. But we *must* all wisely deal with technology in our careers. Engineers especially are expected to be the leading initiators and proponents of developing technology.

One thing that cannot be denied is the ever increasing rate at which technology is expanding. It reaches into almost every aspect of our lives. Technology is the application of science, and science includes everything from the lever and fulcrum to electrons and light beams. One author used a slide rule early in his schooling and remembers his father buying a four-function calculator for $400. We both remember the frequent use of tables to look up trigonometric, logarithmic, and other functions. In the span of a decade they all became obsolete. We recently designed a course to teach basic engineering economics and debated for some time whether to even include the use of tables to look up the time-value-of-money multipliers. For less than $40, a person can buy a calculator that is easy and intuitive to operate and absolutely outperforms the tables.

In an article on computers as the instruments of change, Megan Barkume (1992) draws on two scenes. The first is 1952, and scientists and technicians wait one-and-a-half minutes for a multimillion dollar computer to solve an equation. The second is 1992, and an "ordinary" person sits at a personal computer, which costs a few thousand dollars, and waits less than a second for the solution to the same equation. Barkume says that the capabilities and speed of the microprocessor chips, the heart of today's computers, is doubling every two years while at the same time the price drops.

Computers have changed the way we do business, even the way we think. For most of us, it would be impossible to function without a word processor and spreadsheet. We are all users of databases. It is almost impossible to imagine overheads that are not created from some sort of presentation software. And who hasn't been caught forgetting to use the spell checker?

We try to envision the office environment of the future. Will it be like the "Jetsons" or "Star Trek," or even more incredible? Will it even be an office? In writing his 1952 book, *Player Piano*, Kurt Vonnegut tried to conceive the most elaborate computer he could. He writes of a computer that filled immense caverns, its logic driven by innumerable vacuum tubes. His vision is now archaic.

It is becoming more and more difficult to distinguish scientific fact from fantasy. An issue of *New Scientist* (April 1993) was received by many of its subscribers on April Fool's Day. They opened the magazine to find three articles on incredible subjects: the teleporting of particles at the speed of light, capturing asteroids and using the water they contain to run reactors, and finally an article that described the introduction of genes from sunflowers into clover plants so that the sheep that fed on them would produce more wool. Not to be duped, scores of readers called to let the editors know that they had seen past the ruse, and that these articles had been planted to test their credulity. As it turns out, all of the articles were indeed based on fact. In a comment to their readers, the editors noted, "Arthur C. Clarke (a well known author of science fiction) once pointed out that the products of advanced technology would be indistinguishable from magic to people from less sophisticated cultures. It now seems that the pace of change in science has blurred the line between the possible and the impossible and that magic might turn into plain old science around the next corner."

With respect to technology, the engineer is then given three major challenges: (1) Deal with the changes in technology that impact the way they do their own work, including things like new tools, methods, hardware, software, etc.; (2) Offer input to guide your organization's decisions on directions to take as new technology comes along and then to successfully implement those new systems; and, (3) Use wisdom in meeting the first two challenges.

Keep pace with advances in technology

Currently, the most apparent developments that affect all of us come as a result of dealing with computers, and especially with personal computers. It is critical that the engineer who wants to stay competitive in the work place have a user's knowledge of hardware and software. The interface to mainframe and minicomputers tends to be much less user friendly than the PC, but often they provide the only access to much-needed data.

There are a few types of PC-based software packages that are in common use in most companies. All companies will have at least one word processor package available to its employees as well as one or more spreadsheet programs. Generally, a presentation package will also be in use. Many companies have user-definable databases on their PCs. Realistically, one can't be familiar with all of the software packages available. Instead, one should possess a willingness to explore and become comfortable with the use of reference manuals.

Most computer manuals include a "Quick Start" section, which details the basics of the software in a few pages. Beyond that, being able to research answers to questions through the table of contents and index can make software usable in most cases. Also, software packages are beginning to take on a similar look, using the same key strokes to perform similar functions, which is helpful in learning new software.

But computers are certainly not the only technology in the work place. As a business community, we are still trying to define the proper use of phone mail and automated answering and call-handling systems. We are surrounded by technology and most of it is changing. (The technology of staplers and tape dispensers seems to be fairly stable, but just about everything else seems pretty dynamic.)

As the use of electronic calculators has gone from unimaginable to commonplace in the last twenty-five years, imagine an engineer that refused to give up math tables and the slide rule. Failing to make that change, it would have been even less likely that he or she adopt use of the personal computer, especially since its applications in the beginning were fairly limited. Imagine this engineer's functional disability in the engineering environment of today. More and more, engineers and drafters are working on-line and work is becoming increasingly paperless. Hard as it is to believe, there are technologically impaired engineers in this or similar situations, and their only real option, having given up on change, is to weather it out until retirement.

It is easy to have the mistaken belief that because one understands and is comfortable with today's technologies that they are safe. Twenty-five years ago, our "behind-the-times" engineer was up-to-date. Three things seem certain: technology will rapidly change; technology will always be at the center of our work processes; and, finally, one can either keep up or be left behind.

The need to keep up to date is complicated by the great breadth of the technological front. There is no way to have even a working understanding of all the machines around us. Our advice is to stay current with the basic technologies, especially the use of personal computers, networks, modems, etc. We feel the client-server environment, using multiple products from multiple vendors, will be used in business for some time to come. The notion of a data warehouse, whereby all sorts of data are stored centrally, typically in relational databases, and accessed by users throughout the company as they need it, will also be a part of our work into the future. Being comfortable with these basic technologies is important. Telecommunications will be at the center of work life, and so a knowledge of advances in this area will be important. There are also basic technologies in every engineering field.

Access to other technologies can be maintained by developing a network of others who are familiar with them. Often, these networks are built into the corporate organization. Be certain to take advantage of the abilities of those with whom you work. One of the best means we have found to stay current with changes in our field has been active involvement with professional societies that provide plant tours and a time to network with others in related occupations. Ask questions and solicit help, and through them keep abreast of useful technologic applications.

Provide input to guide the organization

When organizations are making decisions related to technology, it is common for them to call upon their engineering staff to provide a technical perspective and economic reasoning. When researching the merits of adding equipment, software, or other technology to a company, a number of points should be considered.

Take as broad a look at solutions as you can. Trade magazines, vendors, other companies, and professional colleagues will provide several sources of information. Keep as open a mind as possible to look at non-intuitive alternatives. For your own peace of mind when researching technology, make certain the vendors understand you are strictly involved in fact finding and data gathering, and don't be intimidated by their desire to make a sale. Take advantage of the networking you have done to see how other companies have addressed the problem. Check with others in your company and in other companies to see how the problem was dealt with in the past and what complications arose.

Check the references provided by the vendors. Even though these references have been handpicked because they are favorable to the product, they will generally have a much clearer perspective of the merits of the product. Ask them a lot of questions, not only regarding performance, but also the frequency and cost of breakdowns, the need and cost for training of operators, the cost of remodeling to accommodate new equipment, etc. Two important things to find out are the reasons they bought the item (what problem they were trying to solve), and whether they think the item will solve your problem. A real hazard in investigating new applications for a company is getting so caught up in the possibilities of the solution that one loses sight of the problem being solved.

Also check on the need for other equipment that is required for operation of the item you are considering. Ask if there were dramatic changes in power consumption or problems with hazardous waste that was generated as a by-product of the process. Obviously, it is possible that you may miss some of the critical questions. Consulting with coworkers and other associates in your network will help you come up with the list of most crucial considerations. It will probably take some time to conduct these interviews with your references, but the time spent up front will be well worth it. If you are able to perform this research at their place of business and get a demonstration of the equipment in place, so much the better.

Always be skeptical of the promises made by a potential supplier of technology. Their enthusiasm for their product and optimism that it will perform in your situation will not carry you through the terrible moment you realize your company has spent a considerable amount of money for equipment that may not fit its needs. An example follows.

> An engineer was given the responsibility of researching and recommending a new telephone system for his company. The president of the company had a penchant for gadgets, and wanted a system that would provide good service for several years, and also included an automated attendant and voice mail. Several systems were researched (all of which *looked* good), a recommendation was made, and a deal was struck.
>
> It took the installers longer than expected to convert to the new system, in fact from early Friday evening and into Saturday morning, but finally

everything appeared to be working. And it did work...until mid-morning on Monday. This was a business that survived as much as any other on phone communications, and one by one the telephone lines locked up, preventing both incoming and outgoing calls. The agony of that day and the weeks that followed are clearly etched in the engineer's memory.

As it turned out, the company had successfully installed phone systems, automated attendants, and voice mail, but never all three in the same facility. There was a bug that made two of the modules incompatible with the third. The experience not only cost this young engineer some of his credibility, but also marked the growth of his first gray hairs.

It is important to develop trusting alliances with vendors, but it is also important to realize that you often do not share the same goals, and they can in no way take over your responsibility to look out for the best interests of your organization. In the same line, it is not advisable to make purchasing decisions based solely on your rapport with the vendor rather than on the performance of their product.

Consider the options. When automation of a process is being considered, it may be wise to first consider other, more economical and easily realized savings. Look for ways that the work station can be redesigned and the work methods improved. It is often the case that the labor savings promised from automation actually come from the change in methods that is required to implement the new technology, and all or part of these savings could be realized without the machine. The process of making several gradual, relatively small changes is often more effective than the more drastic reengineering that is usually required by automation. It may be advisable to first update layouts and methods and then look at updating the associated technology.

Consider the costs. As an engineer, it is especially important to anticipate *all* of the costs. Not only are there the obvious costs of purchase, shipping, installation, and modifications to existing facilities or construction of new ones, but there are also costs of training, costs of interruption of processes, costs of operation, etc. There may also be changes in licensing or certification requirements of the machine or operators. Intangible considerations might include impact on morale or impact on the perception of your company in the community. Other companies who have already installed the item being considered are a very useful source of this information.

It is important to isolate the sunk costs (financial decisions that have already been made and the resources already spent) of the current systems. Although fellow employees will probably have strong emotional attachment to these sunk costs, these feelings should not enter into the cost considerations of purchasing the new system.

Implementing new technology

If you find yourself responsible for the implementation of a new system, the following points are useful considerations.

Make certain there is consensus on the implementation. It should be known exactly what is to be purchased and how much it will cost. This includes equipment, materials, labor, etc. Confirm that the funding for the project is indeed available and the expenditure approved. Know what contracts have been signed and the important implications of each. Find out who in the company has responsibility for approval of the installation, maintenance, and operation of the equipment after it is installed.

Develop an implementation plan, especially for more complex systems. If a plan has been established, it will be much easier to get buy-in from those involved. They can see where they are involved and what time demands will be placed on them. A formal plan will also provide a vehicle for feedback, whereby others can see both possibilities for improvement and essential steps that have not been considered. The plan will also help minimize unrealistic expectations related to the schedule of implementation. This is true in general of all plans involving several groups of employees. In one case, a customer had ordered several units of a product that had to be substantially altered to meet their application needs. A delivery commitment was made before a development and production schedule was developed (a common mistake). The delivery date *seemed* like a long way off, so the informal schedule seemed feasible. It wasn't until all departments were brought together and a Gantt chart drawn, that it became obvious that the delivery schedule was unreasonable. Creating a formal plan alleviated a lot of grief in-house and also prevented upsetting the customer when the unrealistic date passed. The formal plan allowed a reasonable negotiation between internal departments and the customer.

Implement change one step at a time (also known as "chunking"). This method is almost always superior to drastic change. The problems encountered in the telephone system crisis described in the preceding section would have been mitigated if the components of the phone system had been installed one piece at a time. The incompatibility would have been apparent and the offending system taken off line.

Piecemeal implementation allows employees to adapt gradually, speeding the learning curve and decreasing the impact on operations. Sometimes the step-wise approach is not feasible because of the drastic nature of the change or some other issue. It may be, however, that the system is implemented all at once out of impatience and not for any credible justification.

Keep open lines of communication with everyone involved. If something doesn't perform as expected, don't just live with it, involve the vendor in resolving the problem. If the project falls behind schedule, let those impacted know about the delay and any revisions to the timetable. It is our experience that bad situations become worse when people who should be involved aren't.

Case study: The Boeing 777

An example of the prudent use of leading-edge technology in improving work processes was the development of the Boeing 777. Appropriate technology selection not only allowed them to complete the project in less time and less cost, but radically improved the design process.

In the mid-1980s Boeing found themselves without an aircraft to compete in the growing midsize market, and was losing sales to Airbus A340 and A330, and the McDonnell Douglas MD-11. Rather than redesign the existing 767, Boeing sat out to design and produce the new 777 aircraft in radically new ways. They knew that to be successful in the current marketplace, the aircraft would have to be very cost-efficient to purchase and operate and offer the highest possible quality. (The key measures of quality involve safety, flexibility, and user-friendliness.)

To meet these goals, Boeing pursued a two-part strategy: (1) The 777 was to be the first Boeing aircraft designed entirely on computers, and (2) The design team was to be comprised of all functions related to the aircraft, including not only traditional design groups, but also tool design, suppliers, material, production, marketing, sales, and customers.

The computer system revolved around eight of IBM's most powerful mainframe computers (the ES/9000-900) serving about 2,000 terminals, and used CATIA software designed by Dassault Systems (Dornheim 1991). This design system eliminated the process of translating three-dimensional concepts to blueprints, a two-dimensional medium, only to be translated again to the three-dimensional mock-ups and models. In fact, the mock-ups and the plaster master model were eliminated. The computer allowed the design engineers to quickly view a single part from any angle as well as view the part as it fits into the whole. Effects of changes to one part on related pieces were seen and dealt with immediately.

This digital model is expected to be five to ten times faster than the master model method of blueprints, mock-ups and plaster models (O'Lone 1992). This "paperless" system has allowed Japanese partners, using high-speed networks, to work on subassemblies in real time as if they were at the Washington state facility (Cook 1994).

Another feature of the computer system was "CATIA-man," a virtual airline mechanic used by designers to conceptually test the serviceability of the aircraft. For instance, it was found that the mechanic could not reach the red navigation light on the 777's roof to change it, so the light was moved long before production ever began (Stover 1994). CATIA-man was used to test other ergonomic considerations throughout the design.

At the end of each stage of design, software was used to make a complete review of the structures for interferences, consider the nature of tooling required, and evaluate producibility in general, processes that too often in the past had taken place on the production floor.

The net result of using this three-dimensional interactive design was to reduce rework and changes in production by 50 percent when compared to the 767. The changes that were needed were accomplished faster and resulted in fewer ripples downstream because of the comprehensive view offered by this technology. Another benefit to production was the greater degree of precision in the parts being assembled with fewer calls for "a bigger hammer" to fit pieces together (Proctor 1993).

Wisdom and technology

Wisdom is good judgment. A favorite quote from an unknown source is, "Good judgment comes from experience. Experience comes from bad judgment." Through planning and seeking the help and cooperation of others, it is hoped that experience can also be gained through less painful means. There are some important lessons to be learned in the application of technology.

Like the sirens of Greek mythology whose songs lured sailors to their deaths, technology often has the effect of distracting us from the real business at hand. Attention is often diverted from what is truly important, meeting the needs of customers for instance, to things that are less important, say, supersystems that are meant to help employees meet the needs of customers and may or may not hit the mark.

The importance of technology has been stressed and its benefit to companies explained, but it remains the case that sometimes a stated goal of improved customer service becomes overshadowed by the pursuit of systems. For example, a friend was telling us of a coworker that spent several hours refining a spreadsheet that would calculate the answer to a problem that could have been solved by hand in a few minutes. The reasoning was that the spreadsheet could be used to repeat the calculation for similar problems in the future. Unfortunately, there was actually a very limited demand for the calculation but the worker's enthusiasm for solving the problem with technology eclipsed the logical and efficient use of a limited resource: the employee's time.

We wish we could say this has never happened to us. We are familiar with one company that had an incredibly intricate system for collecting the time spent by production workers on each *piece* of their product. The system was so complex and the amount of data so great that a department of several people was required to maintain it. It was also so complex and large that retrieving any useful information was left to a few ambitious interns. Other than regular reports that were distributed and ignored, we could see no useful purpose. Our experience would suggest that wisdom is setting a goal and keeping a clear focus on that goal as a technological application is pursued.

We would certainly not suggest that the answer to technology is always "no," or that cheaper and simpler is better. One of the reasons that People's Express disappeared so quickly after a phenomenal entrance in the passenger airline industry was their reluctance to join the other airlines in developing a computerized ticketing system.

Good judgment is the efficient use of limited resources, typically time, money, and the trust and commitment of employees to meet the needs of customers. Remember, efficiency is defined to be the reduction of inputs and the maximization of outputs. Why buy all the bells and whistles when the half-life of a technology is probably a few years, and the total life in the company may be less than ten? Bigger isn't necessarily better, it's just bigger. Work to solve the immediate problem with an eye to the future. Look for applications that are expandable. This is easily said, but not easily accomplished. Again, we think it can be done through an absolute focus on a clear goal and with the involvement of others.

Finally, change is almost certain to be met with suspicion. This suspicion usually comes from fear: fear of job loss, lost importance, lost involvement, etc. These fears may, of course, be rational, as may the fear that the solution chosen could be detrimental rather than beneficial to the organization. One of the main tasks of a proponent of change is to alleviate fear. It is an unreasonable expectation that new technologies will be embraced in application. Be prepared to present a convincing argument.

Steven Hronec (1993) provides an insightful presentation of

handling change within an organization in his book, *Vital Signs*. A president of a company shared a story with him about the repercussions of a change in his sales organization. The first thing one morning he held a meeting with the eight regional sales managers and explained a new policy on the payment of commissions. The commission rate and the number of product lines were to be increased, and the commission checks were to be processed in half the time it had taken in the past. Within a few hours he had received about 75 phone calls from sales people asking why he was changing the policy. He was incredulous that despite the strictly positive change, the sales force had trouble accepting it. Hronec suggests four "enablers" that help mitigate the negative impact of positive change. They are communication, training, rewards, and benchmarking.

We have discussed the importance of communication, the two-way exchange that promotes understanding, refining, and buy-in. When those involved in change are brought on board early in the process, informed of possible directions, and offered a means to contribute ideas and suggestions, they will be supportive of the change when it is time to put it in place.

Well-prepared training, targeted to meet the employees' needs, helps them feel comfortable in a changing environment. Employees can be rewarded for their contribution to the group with anything that is of value to them, not just money, but also public recognition and praise. Rewards help keep the group focused on the end result during the implementation phase, when it is often difficult to believe that the pain of change is worth the seemingly uncertain outcome. Finally, Hronec says that benchmarking allows companies to see their progress in relation to their peers, competition, customer expectations, and their former selves.

In conclusion, our suggestion would be that the use of technology is critical and inevitable, and should be employed to find the best possible solutions to problems. The spotlight should always be on solving the problem, not on appealing gadgets.

Chapter 5

Finance:
The Bottom Line

We have talked extensively about holistic engineers, customer satisfaction, getting everyone involved, and open decision making within organizations. We have stated that businesses are looking for engineers who can work in teams and help bring out the best in everyone. We have mentioned the movement of the business world toward an empowering and coaching style of management and away from command and control. The question may be asked as to why these changes are taking place. We would like to suggest that it has been due to necessity and an act of survival. Businesses around the globe that have been using these principles have been taking market share and profits away from those that haven't been using them. In the first three years of the nineties the top three U.S. auto makers lost over 53 billion dollars. These industrial giants are probably among the few companies who could sustain such losses and not go bankrupt.

The issue is survival in a global marketplace. In the economic battlefield, success is measured by net income, strength of balance sheets, and cash flow. Organizations exist because they generate wealth for their owners by providing a service or a product for which consumers are willing to pay. The ownership can range from an individual proprietor to a very large group of people who hold stock in a company. At either end of this spectrum, owners expect a return on their investment commensurate with the risks they have taken.

Customer satisfaction has become a central issue for many corporations because it has a strong and direct correlation with long term profitability. Companies that are well known for their outstanding customer service have sales and income goals right along with customer satisfaction goals. Any individual, not just an engineer, who wishes to be a major contributor to an organization needs to have a good understanding of the finances of their company. This understanding can be achieved by learning about the income statement, balance sheet, statement of cash flow, and the budgeting process. The mastery of engineering economics is not enough, as crucial as this skill is. We feel engineering schools do a fine job in equipping their graduates with the latter skill and in the context of this chapter we are going to address the former issues.

Balance sheet

The balance sheet is the financial statement that provides information about the assets of an enterprise and how they are financed at a particular point in time. It is called a balance sheet because the financing must be equal to the assets.

$$Assets = Financing$$

Financing is usually composed of two parts, liabilities and shareholder or owner's equity. There are the rare instances where a company is financed solely by owner equity. Even more rare are companies that are financed entirely by debt. Generally, the most economical approach is a combination of the two as we will briefly discuss in the section on capital budgets. The foundation and basic formula for a balance sheet is:

$$Assets = Liabilities + Shareholder Equity$$

The assets and liabilities are divided in two parts, current and noncurrent. Current assets include cash and those assets that are expected to be converted to cash, sold, or used within approximately one year from the date of the balance sheet (Table 5.1). Current liabilities are the debts that are expected to be repaid within a year (Table 5.2).

Included in current liabilities are items such as accounts payable, salaries owed to employees for work that has been performed but not yet

Assets	
Current	Noncurrent
Cash	Land
Accounts Receivable	Equipment
Inventories	Buildings
Certificates of deposit	Patents

Table 5.1. Assets.

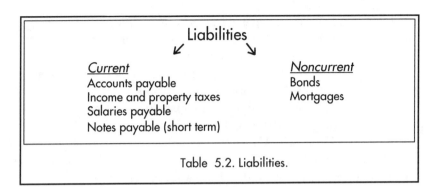

Table 5.2. Liabilities.

paid, and property and income taxes owed to the government.

Noncurrent assets are usually held and used for several years (Table 5.1). Examples of long term assets are buildings, patents, land, and equipment used in producing or providing goods and services to customers. Noncurrent liabilities are the long term debts such as bonds issued by a company, or ten, twenty, or thirty year notes borrowed from a bank or insurance company (Table 5.2). Along with shareholder equity, noncurrent liabilities are an enterprise's long term source of funds.

A simple case study

Lawrence, having worked for a large corporation for a few years, decides he is going to strike out on his own and start an engineering consulting firm by the name of *Quality Consulting Services*, a sole proprietorship. He decides to invest $20,000 he has in savings into the business.

At this point his balance sheet is very simple, as illustrated in Table

QUALITY CONSULTING SERVICES

Balance Sheet (Two Months Before Start of Business)
March 1, 19__

Assets		*Liabilities & Shareholder Equity*	
Cash	$20,00	Liabilities	$0
		Shareholder/Owner Equity	$20,000
		TOTAL LIABILITY AND	
TOTAL		SHAREHOLDERS	
ASSETS	$20,000	EQUITY	$20,000

Table 5.3

5.3. Lawrence believes he needs at least $30,000 to get started correctly. He needs a computer, a CAD/CAM system, and some furniture for his office.

For the additional $10,000, Lawrence decides to go to his friendly neighborhood banker. After a number of visits and in-depth discussions about the potential of his business, anticipated income, and cash flow over the next few years, the bank agrees to lend Lawrence $10,000 on two notes of $5,000 each. One note will be due at the end of one year and the other at the end of three years at a specified interest rate (12% on a $5,000 one-year note and 11% on a $5,000 three-year note).

After receiving the money into his account, Lawrence decides to update his balance sheet. The result of his work is displayed in Table 5.4. Notice that cash and total assets increased by $10,000 and total liabilities and shareholder equity also increased by $10,000. Shareholder equity stayed the same while current and noncurrent liabilities each went up by $5,000. The statement is in balance.

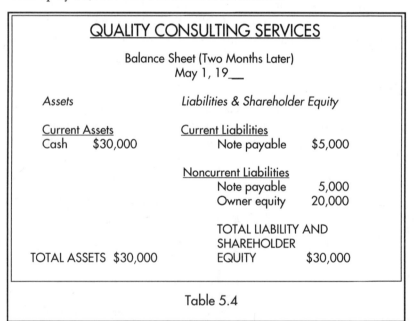

QUALITY CONSULTING SERVICES

Balance Sheet (Two Months Later)
May 1, 19___

Assets	Liabilities & Shareholder Equity	
Current Assets	**Current Liabilities**	
Cash $30,000	Note payable	$5,000
	Noncurrent Liabilities	
	Note payable	5,000
	Owner equity	20,000
	TOTAL LIABILITY AND	
	SHAREHOLDER	
TOTAL ASSETS $30,000	EQUITY	$30,000

Table 5.4

His business account loaded with cash, Lawrence goes shopping for the items he needs. He pays cash for the computer and the CAD/CAM system. For the furniture however, he takes advantage of a 90 day same-as-cash deal offered by the furniture store. Wanting to have a thorough understanding of the business side of things he decided to update his balance sheet (Table 5.5) even though only a week had gone by since the last time he prepared it.

QUALITY CONSULTING SERVICES

Balance Sheet
Start of First Year of Business

Assets	Liabilities & Shareholder Equity	
Current Assets	**Current Liabilities**	
Cash $8,000	Note payable	$5,000
	Accounts Payable	1,500
Noncurrent Assets	**Noncurrent Liabilities**	
Equipment 22,000	Note payable	5,000
Furniture 1,500	Owner equity	20,000
	TOTAL LIABILITY AND SHAREHOLDER	
TOTAL ASSETS $31,500	EQUITY	$31,500

Table 5.5

Cash has been reduced by the $22,000 he spent on equipment, which now shows up on the asset side of the balance sheet at its acquisition cost of $22,000. Total assets increased by $1,500, which shows up as furniture under noncurrent assets. To bring the balance sheet into balance Lawrence lists the 90 day payment for the furniture as accounts payable under current liabilities (since it has to be repaid before one year has passed) in the amount of $1,500, thereby increasing total liability and owner equity to $31,500. Lawrence can now see that a balance sheet reflects the financial position of a firm at a given moment in time and it changes as each transaction is recorded.

In addition, each transaction is an exchange and the two sides of the transaction are always accounted for. For example, when the furniture was bought on credit, assets increased and so did the liabilities. And when, at the end of 90 days, the furniture bill gets paid, cash will be reduced by $1,500 and the accounts payable will vanish, reducing both the "assets" and the "liabilities and owner equity" sides of the equation down to $30,000. Owner equity can be increased by making a profit and is decreased when cash withdrawals are made by the owner.

Income statement

The income statement is the second main financial statement. It reflects the results of the operation of an enterprise for a period of time.

The basic formula is:

Revenue - Expenses = Net Income

Net income is used interchangeably with profits and earnings, depending on which text or annual report you are reviewing. Most enterprises are created and conduct business with the goal of generating a profit. The income statement is the measurement tool for determining how successful a firm has been in accomplishing this goal over a given period of time. According to a financial accounting text (Davidson, Sticknay, and Weil 1988): "Revenues measure the inflow of assets (or reductions in liabilities) from selling goods and providing services to customers. Expenses measure the outflow of assets (or increases in liabilities) used up in generating revenues. For each expense either an asset decreases or a liability increases."

Let's return to our simple case and see how Lawrence's firm has performed during its first year of operation.

Consulting fees during the first year were $89,000, which essentially made up all revenue. Expenses are listed in Table 5.6. Clerical support was the highest expense followed by office space rent. The next highest expense however was not a cash item. Lawrence did not have to pay anyone $4,700 for depreciation. This is the amount the assets of the firm lost in book value or an estimate of what portion of noncurrent

---

QUALITY CONSULTING SERVICES

Income Statement for Year One

Revenues:	
Consulting fees	$89,000
Interest on checking account	250
Total Revenues	89,250
Expenses:	
Clerical expenses	25,000
Supplies	2,500
Office space rent	12,000
Interest expense	1,150
Depreciation	<u>4,700</u>
Total Expenses	45,350
NET INCOME	$43,900

Table 5.6

assets were used up during the first year of business. It is because of items such as depreciation that net income is not the same as cash from operations. The cash flow statement will further explain this issue in the next section.

Net income increases owner's equity and any withdrawals reduce it. For example, let's assume that Lawrence's old job paid him $35,000 per year. Since he does not receive a salary from his own firm, he has to make cash withdrawals to pay for his living expenses. Another way to think of these withdrawals is to consider them a distribution of net income. In large corporations with many owners or stockholders this distribution is called dividends. The withdrawal of cash is reflected in the balance sheet. Table 5.7 shows *Quality Consulting Services'* balance sheet at the end of the first year of doing business. The first $5,000 note has been paid off. Net income has been added to owner's equity and Lawrence's cash withdrawals have been subtracted. Depreciation for the equipment and furniture is reflected on the asset side.

QUALITY CONSULTING SERVICES

Balance Sheet / End of Year (1)

Assets			Liabilities and Owner Equity		
Current Assets			**Current Liabilities**		
Cash		$16,900	Note payable	$	0
			Account payable		1,800
Non-Current Assets			**Non-Current Liabilities**		
Equipment	22,000		Note payable		5,000
(less depreciation)	(4,400)				
Furniture	1,500		Total Liabilities		6,800
(less depreciation)	(300)				
			Owner Equity		
Net Non-Current	18,800		Owner equity on day one		20,000
Assets			Add net income		43,900
			Less withdrawals		(35,000)
			Owner Equity		28,900
TOTAL ASSETS	$ 35,700		TOTAL LIABILITIES & OWNER EQUITY		$ 35,700

Table 5.7

For the purposes of this example, both equipment and furniture were depreciated over a five year period. Earlier it was stated depreciation is an estimate of what portion of noncurrent assets was consumed during a particular period of doing business. In the case of furniture, obviously by the end of five years the furniture will not be useless and Lawrence, most probably, will not have to replace it. The point here is that the book life of an asset may be, and usually is, different from its real life. The reason for this discrepancy is usually related to tax laws, a discussion of which is beyond the scope of this book. Lawrence's consulting firm had a great first year. Most new ventures take a few years before they become profitable.

Cash flow statement

The cash flow statement is the third principal financial statement. It

---

QUALITY CONSULTING SERVICES

Cash Flow Statement
First Year of Operation

SOURCES OF CASH

From Operations:	Net income	$43,900
	Plus depreciation	4,700
	Plus increase in accounts payable	300

Total cash from operations	48,900
Less withdrawal by owner	(35,000)
Cash reinvested in firm	13,900
From financing: Bank note repaid	(5,000)
Total Sources of Cash	$ 8,900

USES OF CASH

Invested in assets	$ 0

Total sources less uses	8,900
Plus cash at beginning of year	8,000
CASH AT YEAR END	$16,900

Table 5.8

provides information about sources and uses of cash in a business. Sources of cash are divided into two groups, internal sources and external sources. The internal source is essentially profits of the company. Selling of assets is also a source of cash. External sources for funds are loans from banks or lending institutions and owner's capital.

Let's examine the statement of cash flow for *Quality Consulting Services* for its first year of operation (Table 5.8). The main source of cash was the profits the firm earned. To that we add the non-cash items: depreciation and increase in the accounts payable of $300 from the time the business started operating. Depreciation is added back in for cash flow purposes because depreciation is a non-cash accounting means to spread the original cost of the equipment over the several years the equipment will be used. The original cost is counted only once in the cash flow statement (at the time of purchase). It is not counted again when part of it is written off as an expense through depreciation. The increase in accounts payable is added back in for the same reason: No cash has actually changed hands. In calculating net income, $2,500 was subtracted for supplies, some of which had not been paid for and is reflected in the increase to accounts payable.

Cash from operations, therefore, was $48,900 from which Lawrence's $35,000 withdrawal was subtracted, leaving $13,900 for reinvestment in the firm. The one-year $5,000 note was paid off and shows up as a negative financing activity and reduces the amount of cash to $8,900. During the first year Lawrence did not make any investment in assets for the business, explaining the zero for the "Uses of Cash" section. Adding the cash at the beginning of the period, the "Cash at Year End" figure is the same as the "Cash" amount on the balance sheet for the end of the first year. The cash flow statement illustrates how the beginning cash amount changed into the ending cash amount.

The case of *Quality Consulting Services* is a simple one, yet useful for illustration purposes. In the business world and especially in large corporations, balance sheets, cash flow, and income statements get significantly more complicated but the principles are the same. Our recommendation to engineers is to get to know the balance sheet and income statements of their company and track it over time.

Budgets, budgets, budgets

A few years back, our company was having a lean year and revenues were not at the level we had anticipated. A supervisor who had a particular project postponed was complaining about the postponement and he did not understand why the dollars were not available when he had budgeted for them over a year ago. The issue this supervisor had not realized is that most, if not all, businesses who prepare budgets do not have the luxury of sticking to their budgets, regardless of what happens in the marketplace. Budgets are reviewed and approved with certain assumptions regarding level of sales, production, cash flow, etc. If any of these variables do not materialize as expected, compensating adjustments need to be made in other variables to minimize the impact on net income.

But why is net income so important? Why do businesses cut expenses and reduce their workforce to protect their income? Why aren't companies just happy with a lower level of income instead of laying people off? The answer is relatively simple if you bring the question to a personal level. Let's say you get a letter in today's mail from your bank that states they are having a bad year and instead of paying 5 percent interest on a passbook savings account, they are going to pay 1 percent interest over the next year because they want to keep everyone employed and avoid the hardship of cutting budgets any more than they have to. The letter closes asking for your understanding and thanking you for your business. Would you leave your savings in this bank or move it to the bank across the street that is continuing to pay 5 percent? Most people would move their account and that is why you have not received, nor are you likely to receive, a letter like this in the future. All for-profit organizations (as opposed to nonprofit) are in business to generate wealth for their owners through profits, whether they be sole proprietors or a large group of stockholders.

In the case of *Quality Consulting Services*, Lawrence needed to have a net income of around $30,000 to pay his living expenses. Looking at his income statement, what would you have done if you were in his shoes and consulting fees were one-half of the amount shown? In large corporations the same philosophy exists but on a much larger scale. The management of the company, who are the custodians of the assets, do not know all the owners (i.e., stockholders), but their commitment is, or should be, just as strong to provide the owners a fair return on their investment.

Operating budget

A tool used to help ensure the enterprise will produce a net income during its next year or period of operation is an operating budget. An operating budget is the best estimate of how business will be conducted during an upcoming period, usually a year. Revenues and expenses are estimated based on historical data, forecasts of economic and industry-specific trends, as well as management knowledge and wisdom. Using this information the income statement is developed for the coming

QUALITY CONSULTING SERVICES					
Budget Comparison Report					
	Year 1 Actual	Year 2 Actual	Year 3 Budget	Year 3 Actual	Year 4 Budget	Year 3 vs. Year 4 (%)
Revenues:						
Consulting Fees	$89,000	$65,500	$80,000	$92,000	$90,000	- 2%
interest	250	120	140	160	170	6%
Total Revenues	$89,250	$65,620	$80,140	$92,160	$90,170	- 2.0%
Expenses:						
Clerical	$25,000	$9,200	$29,500	$28,000	$26,000	-7%
Supplies	2,500	1,800	2,100	2,800	2,300	-18%
Office Space Rent	12,000	13,200	13,860	13,860	14,453	5%
Interest	1,150	550	550	550	0	-
Depreciation	4,700	4,700	4,700	4,700	4,700	0%
Total Expenses	43,350	29,450	50,710	49,910	47,553	-5%
NET INCOME	$43,900	$36,170	$29,430	$42,250	$42,617	1%

Table 5.9

year. The accounting term for a future period income-statement is a pro-forma income statement. Operating budgets in small firms are simple and easy to construct. Table 5.9 is a budget comparison report for *Quality Consulting Services* for its fourth year of operation. Usually, along with the next year's budget information, historical data associated with each budget item is also reported to illustrate trends that may exist.

This year's budget is compared with last year's budget, as well as with the actual results achieved. The last column in Table 5.9 displays the difference (in percent) between the year 4 budget and the year 3 actual results. All the information in a budget comparison report is intended to help the decision makers come up with the best estimate of the coming year's results by questioning all the assumptions. For example, in the case of *Quality Consulting Services*, is it reasonable to expect that supplies expenses will be reduced by 18% when revenues are anticipated to decrease by only 2%? What about the 7% decrease in clerical expenses? Lawrence has found a new source for the supplies and believes the 18% reduction is realistic as well as the decrease in clerical expenses due to automation of some tasks.

Even with this simple example where all the calculations could have been performed in Lawrence's mind, preparing a budget is a helpful and useful process. Imagine how valuable it becomes in larger organizations with numerous budget centers. On this larger scale, the data from all the various departments has to be rolled up to develop a pro-forma income statement. Once developed, the income is usually not at levels acceptable to management because the expenses often include items from everyone's wish list.

Depending on the organization, one of two approaches is pursued at this point to bring the net income picture back in line. The easy approach is to ask all budget centers to cut their expenses by a percentage which would bring the earnings up to the level anticipated by stock analysts and shareholders of the company. The more difficult approach is to analyze each individual budget and make substantial cuts in those heavy with wish-list items.

Another criteria used in the analysis could be impact on customer satisfaction. The latter process is more time consuming, but it has the advantage of reducing costs associated with low value-added activities while maintaining or increasing the investment in high value-added initiatives. The first approach asks for certain percentage reduction in all activities regardless of their added value.

In large organizations, the budgeting process for any particular year starts as much as six to nine months in advance. In the absence of data for a full past year of operation (since there could be up to nine months left in the current year), most firms provide actual data on the most recent twelve-month period. The upcoming year's budget is compared to the actuals obtained in the most recent 12 months, as well as the current year's budget.

An example of the budgeting process

Lauren, the supervisor of the engineering department in a large manufacturing company, is preparing a budget in June of 1998 for the 1999 operating year. Her statement from the budget department includes data listed in Table 5.10 and she is required to submit her budget for 1999 by July of the current year (1998).

XYZ Company Engineering Department
1999 Annual Budget Preliminary

Year	1995 Actual	1996 Actual	1997 Actual	1998 Budget	June 1997 to June 1998 Actual	1999 Budget
Budget Salaries Travel Training Meals Sundries OfficeSupplies Consultants Utilities • • •						

Table 5.10

Lauren diligently prepares her budget and submits it a week before the deadline. Toward the end of July she receives documents from the budget department stating all departments need to reduce their expenses by 10 percent. Dismayed at the prospect of having to redo her budget, she calls the supervisor of the planning and budget department, Danielle, for an appointment to find a better way to accomplish the budgeting process, without the need for revisiting the budgets once they have been prepared.

At the meeting Lauren asks why the supervisors were not informed if management had a certain target number in mind when they prepared their budget the first time around. Danielle explains that while management has an idea of what the investment community expects from the company in terms of profits and return on equity, they don't know where the revenue and expense figures are going to fall out until all the data from the various departments has been compiled. It is then that projected net income figures can be calculated and compared with the expectations of the shareholders and stock analysts. If the projected net income is below expectations the company has two means to remedy the problem: 1) increase revenues, and/or 2) decrease costs.

Danielle continues, "The economic trends for the upcoming year are such that we believe revenues cannot be increased beyond the levels outlined in our preliminary budget. The only other option available is to reduce cost and that is why you received the memo asking for a 10 percent reduction in operating costs."

This iterative approach to developing an operating budget can be helpful in refining the estimates for the upcoming year of operation. However, there is a point of diminishing returns. Budget managers may become disenchanted if they have to go through the refinement process too many times.

Capital budgets

Capital budgets deal with the investing activities of an organization during its next period of business, usually one year. Businesses determine how much money they will need to fund all of the projects and investments various departments and divisions want to make. Usually there are more projects than there is money to fund them and this is where engineers have traditionally been involved evaluating various capital projects using tools learned in engineering economics classes. Examples of these tools are net present value analysis, internal rate of return analysis, and payback periods. Our intent in including a section on capital budgets is not to review the concepts covered in engineering economics, but to explain how these tools fit into the big picture.

The big picture has to do with how much debt and how much shareholder equity a company has in its capital structure. Capital structure refers to what percentage of the company is financed through borrowing (debt) and what percentage is owner equity. Each industry has its own range as far as reasonable debt-to-equity ratio. One of the major reasons there are varying acceptable ranges for debt-to-equity ratios in different industries is the perceived risk associated with each industry. Industries with high perceived risk, such as technology companies, tend to have less debt, although this is not always the case. Debt is usually preferred over equity since it costs a company less. On the other hand, the interest on debt must be paid regularly regardless of what kind of year the company had, whereas stock dividends do not.

Equity or shareholder investment is more expensive because companies are not obligated to pay dividends to holders of common stock. If the company has a bad year they may choose not to pay dividends to their shareholders in that year and suffer the consequences, which is beyond the scope of this discussion. (There are companies, however, who don't issue dividends or pay very low dividends and the stockholders in these type of companies expect their return in terms of increase in the price of their stock.) Second, stockholders are last in line in terms of claims on the assets of a corporation in case of bankruptcy. That is to say, the creditors, such as suppliers who have not been paid, are the first to get their money when the assets of the organization are sold, then the money left over is divided among debt holders. Debt holders include banks, insurance companies, or individuals who purchased bonds in the company. After all the entities that loaned the bankrupt company money have been satisfied, any funds left over are divided among the owners or stockholders of the enterprise. It is because of this increased risk that investors demand a higher return than those offered to the bondholders of the company.

The question may be asked, "If debt is cheaper why aren't organizations financing their activities entirely with debt?" They probably would if not for the fact that the debt market also evaluates risks and increases the price as the level of debt rises in an organization. Lenders demand higher and higher rates because of the increased probability of not getting their money back. A lender may be willing to loan money at 6 percent interest if he or she has 100% chance of getting the principal and interest back. If the probability is reduced to 90%, some lenders may not want to do business at all while others may charge 18 percent interest per year to bring the expected value of the deal in par with 100% probability at 6 percent. The calculation would be made as follows:

$$100\% \times 1.06 = 1.06$$
$$90\% \times (1 + i) = 1.06$$
$$1 + i = 1.18$$
$$i = 18\%$$

In our personal lives, the same principals hold true. The cheapest form of loan individuals can get is usually a real estate loan, which is secured by the property being purchased. Even then banks and mortgage companies pay close attention to the borrower's source of income. They want the monthly mortgage payment to be no more than 45 percent of the borrower's gross monthly income to minimize the probability of default on the loan. As loan payments become a larger and larger portion of one's monthly income, some institutions refuse to lend money. Banks usually fall in this category. There are secondary markets that are willing to lend money even in these situations but they charge a substantially higher interest rate to compensate them for the increased risk.

The big picture then, in terms of availability of funds for capital projects, depends on a number of factors. First, the company's debt-to-equity ratio and whether it has the flexibility of going to either the debt or equity market to raise funds is a factor. If a company is on the high end of debt-to-equity ratio, it may be forced to issue stock in order to raise the necessary capital. Issuing stock in the equity market is an expensive proposition and is not something companies like to do on a regular basis. If the proposed projects are not lucrative and large enough to justify the expenses of a stock offering, the company may have to forgo the projects when it has used all available credit. Secondly, the conditions in the equity and debt market can have a profound impact on the availability of funds. A company may decide to postpone projects due to unfavorable conditions in either or both of these markets.

For example, if the company's stock is trading at an all-time low, it would not be a good idea to issue stock at that time. If the interest rates are at high levels it may be prudent to wait rather than take out a loan. On the other hand, if interest rates are low it may be time to expedite projects which, at higher interest rates, would be infeasible. The third factor influencing a firm's access to capital markets is its track record in generating wealth for its owners or shareholders. The better the track record, the larger the amount of capital dollars that are available. The assumption of markets is that by providing products and services the customers want and are willing to pay for, the firm will be able to pay off its debts as well as generate a fair return on investment for its owners, and it will continue to do so with additional funds.

In the final analysis, projects with the highest net present value, highest internal rate of return, and shortest payback period will be the ones funded. Engineers need to be aware of the financial condition of the company to better enable them to pursue the type of projects and products that best suit the organization's needs and capabilities at any particular time. This understanding can be gained by studying the company's annual report and it's quarterly report to shareholders. This information is readily available for publicly held and traded organizations. However, in privately held companies the information may not be easily obtained and is probably not available at all.

In publicly traded companies the chairperson's or president's letter to shareholders in the annual report is a good source of information regarding the financial condition of the company, as well as future challenges and opportunities. The annual report also includes the three financial statements reviewed in this chapter. Spending some time with these statements could prove very valuable to engineers looking for ways to contribute to the success of their organization and increase their own value to their employer.

Chapter 6

Data Analysis:
Managing By The Numbers

Did you know that if you wanted to arrange 15 books on a shelf in every possible way, and you made one change every minute, it would take you 2,487,996 years to do it? Or did you know that one pound of iron contains an estimated 4,891,500,000,000,000,000,000,000 atoms and that the earth travels one and a half million miles each day? How about this: if you were to write down in one line on a piece of paper all the different ways the 26 letters of the alphabet could be arranged, you would need a piece of paper 160 million light-years long. Almost one out of every four people on earth is Chinese and one slightly out-of-date estimate says that the U.S. population is collectively carrying about 2,000,000 pounds of fat (Brandreth 1984).

We are absolutely inundated with data. A quick check of the front page of the local newspaper gave twenty-six pieces of statistical data. Even making this observation is itself a statistic. There must be a lot of people who are looking for data, because it seems there are surveys for everything. Right now, there are two surveys sitting on this author's desk waiting to be completed (another statistic). Surveys have become such an art that it would probably be reasonable to do away with elections and rely solely on polls.

For years, decision makers were encouraged to "manage by the numbers." It is still good advice, not just for financial management (see chapter 5), but management in marketing, production, engineering, as well as all other internal support functions. Numbers provide models that enable us to describe, understand, and improve business processes. Of course, not all data is numeric. Descriptors such as color, failure mode, or model name may be very important considerations, but do not lend themselves directly to mathematical operations. These non-numeric attributes are critical but may not be as readily organized and evaluated as numeric descriptors.

Data has many uses in business. In most cases, these numeric descriptors are a necessary requirement in dealing with processes. Benchmarking relies on the use of numeric measures, as do control charts. Financial transactions have an obvious dependence on numbers. The engineering functions would be lost without numeric data.

Use of data

The use of data usually follows four steps: collection, analysis, interpretation, and conclusion. A lack of understanding of techniques of any of the four steps will likely result in the failure of all of them. Unfortunately, with data the failure is seldom apparent and faulty conclusions can at times seem believable.

Collection is the gathering portion, through sampling, direct measurement, surveys, etc. Care should be taken that assumptions regarding samples and populations are well understood, and gathering methods should be statistically sound. It is best if data is collected as close to the source as is possible. It is also important and often difficult to distill the relevant data from all the data that is available. Typically in engineering school, problems contain exactly the right data needed to solve the problem. This is less true in business school case studies, where extraneous or minor data is often included in the problem statement. In real situations, there is no real limit to the number and type of measurements that can be made, or the sources that can be polled and opinions obtained.

One must scrutinize the available data to discern which is truly necessary and sufficient. Again, soliciting the advice of others will protect the analyst from biases and unfamiliarity with the data.

Finally, it is likely that databases exist in several places throughout the business. When employees cannot find the data they need, they will start collecting it themselves, in tally sheets, spreadsheets, databases, etc. This leaves the data gatherer with a few challenges. The first is to identify these data repositories. The work of gathering data is very much like that of an investigative reporter. The facts are seldom readily accessible, and guardians of the facts will frequently be seriously biased and may be reluctant to share what they have. Persistence and trust building are two attributes that will help the analyst in this process. It is also a good idea to corroborate data from more than one source when possible.

Analysis is the refining of the data to provide a useful summary, including statistics such as average, standard deviation, frequency, etc. It is important that the grouping of data is reasonable and included in reporting the results, and that the mathematical operations are done correctly.

Interpretation is the part of the analysis that answers the question, "What do these measurements mean?" Interpretation requires understanding both of statistical methods and the underlying process. These requirements can be met by more than one person working cooperatively.

The conclusion supplies some action to be taken based on the interpretation, "Because of what the measurements say, we should take this action: . . ."

It is unfortunate that these steps are so often misapplied and still remain believable. One author remembers a statistic that was being touted in his driver education course: Eighty percent of all accidents occur within twenty-five miles of the driver's home. The conclusion: Drivers are less cautious when driving on familiar streets and we as driving novices should not fall into this trap. It seems though, that those who drew this conclusion may have made a significant error in the interpretation step because they did not fully understand the process. In a less scientific study, several of us drew a circle with a 25-mile diameter, centered at our homes, on a local map. We decided that at least ninety percent of the driving our families did was within the circle. If that were actually true, we would draw a conclusion opposite to the one presented in the study: Given the percentage of driving that is done close to home, it is actually safer to drive within twenty-five miles of home.

In another business situation, a fairly extensive analysis was done on the external failure of a product with important decisions based on the conclusions drawn. As it turned out though, the computer routine that accumulated information in a database had been misprogrammed and the numbers used in the analysis were not at all reflective of the actual situation. But based on the credibility of the analysis, interpretation, and conclusion, the decisions seemed reasonable.

There are some ways to help avoid being misled by erroneous conclusions. It is always useful to understand the process behind the numbers. If you don't understand it, it is essential to consult with someone who does.

Determine how the data was collected. For instance, a survey was conducted by a pizza delivery business to measure the level of service. To obtain a sample, the service providers were instructed to give mail-response cards to their customers, who in turn were asked to complete and return the cards. An obvious problem was that the service providers were inclined to give the cards only to customers who had enjoyed a positive experience. Regardless of the quality of analysis and interpretation, only one possible conclusion could be drawn from this study: Nothing!

If the collection of data is reasonable, one then needs to know how the calculations were made, particularly actions taken on things such as missing data points and data points that were unusually high or low (outliers). One also needs to know the assumptions that were made in the interpretation phase. In most analyses, assumptions must be made, and they become an important part of the conclusion. Finally, ask yourself and others that understand the process if the conclusion is credible. A healthy dose of friendly skepticism is wise when presented with data.

The data analysis model

The following steps are based loosely on the "Plan-Do-Check-Act" cycle, and provide a map for solving problems and improving processes through the use of descriptive data.

Step 1 - determine what data is needed. What measurements will best describe the process under consideration? The measurements should be as close to the process in time and location as possible and still give a complete enough picture. For instance, if one wished to measure how cold it was in Minnesota in the winter, one could count the layers of coats and sweaters worn by passersby. On the other hand, it would probably be more practical to make a direct measurement of the temperature with a thermometer, rather than the reaction of people to the temperature. As a more practical example, suppose the efficiency of a certain production process was to be determined. An easy measurement would be the number of units completed and passed to the next process. This number would be available from production records. However, measuring the time each unit was completed would provide much more information about the process. It would be possible to see fluctuations in run time throughout the day, between shifts, operators, or raw materials used. Measurements should give the accuracy necessary to make decisions, understanding that there is usually a trade-off between accuracy and cost.

Step 2 - collect the data. Perhaps the data is already available, either in hard-copy or in a computer database, and work can begin on the next step. However it is often necessary to find the custodian of the data within the company. In smaller organizations, this is usually straightforward, but in larger ones it can be very difficult to track down the person with the needed data.

If the data is not available, take the measurements, using proper sampling techniques where appropriate. This can often be tedious and a drain on the time of those who are taking and recording the measurements. Even though data collection may be tedious, it is extremely important that the data is accurate and meaningful. The adage "garbage in-garbage out" certainly applies if this step is not completed correctly. Great care must be taken to ensure the usefulness of subsequent work. It will help if the data collectors understand the importance of the data and how it will be used.

Step 3 - analyze the data. Besides standard statistics such as means and standard deviations and the more sophisticated statistical tools, all of which can be applied to numerical data, there are other ways to give meaning to the gathered information. A Pareto Chart, Run Chart, and Scatter Diagram are techniques that organize and give meaning to numerical data. The Affinity Diagram and Cause and Effect Diagram are examples of techniques that group and organize non-numerical information into logical relationships, giving a much clearer picture of the whole. Several of these tools are described at the end of this chapter. One should seek understanding of the process being studied: what makes the process work, what causes it to break down, what are the measures that best predict the process' success, etc.? It is advisable to take frequent "reality checks" with people who understand the processes being studied to make certain that the analysis makes sense in the real world.

Step 4 - from the analyses of the previous step, develop a plan of action. For the issue being considered, focus on the most significant of the root causes identified in the data analysis. It is unrealistic to work on all problems at once. Choose a few causes and develop remedies that will address them. The success of this process relies much more on persistent repetition of the steps, each time working on a few problems, than on uncovering and developing a single comprehensive solution to all problems. Formulate a solution that identifies steps to be taken, people responsible for each step, time frames for completion of each step, and a measurable goal. Each of these factors is important. If the solution steps are identified, the work becomes much more concrete and progress much more readily seen. This is critical in preventing the discouragement and loss of energy that comes when no progress is perceived. When people are made accountable for a step, it becomes possible for them to take responsibility for its conclusion. Otherwise a team ends up with a lot more confusion than accomplishment. Providing the accountable person with a deadline, and with them monitoring their progress towards accomplishment of the task, allows them an opportunity to prioritize the task with their other work, get help as needed, and be comfortable with the expectations of others in their group.

Having a measurable goal means the task can be accomplished. Without a goal, there tends to be no boundaries to the project and little sense of accomplishment. Orphaned pet projects are adopted, perfection becomes the expectation, and the project often becomes somewhat chameleon, changing in appearance to reflect the current direction. There is also a real danger that the project team may become more of a standing committee, generating busy work without real accomplishment.

The more critical the process is to the core function of the business, and the more complex its solution, the more useful a formal plan of action becomes. A Gantt or Program Evaluation Research Technique (PERT) chart that includes assignments made and a formal goal statement will provide the "handle" by which the task can be controlled.

Step 5 - take action. Team members are often anxious to take action from the beginning and must be constrained from doing so to provide for an adequate analysis and development of the solution. The plan of action produced in the previous step provides and efficient and effective way to actually "do" something about the issue at hand.

The team leader takes on the important role of facilitating team members in their assignments, making sure they have the necessary encouragement, human and financial resources, information, and training they need to accomplish their tasks. Because of complexity of processes and the general difficulty of bringing about change, members will probably need frequent encouragement. This should include periodic reviews of their tasks to make sure progress is in line with the intended target and that deadlines will be met. Leaving members alone to "operate in a vacuum" will jeopardize their personal success as well as the success of the team in general.

Step 6 - collect data. This is a repeat of step 2. Data should be gathered in the same way from the same source to allow a valid comparison in the next step.

Step 7 - analyze the change in the process. Using the data collected in steps 2 and 6, determine the degree to which the process actually improved. Care should be taken not to distort the data to give the appearance of improvement. If the process did indeed improve, repeat the steps beginning with step 1. If the process did not improve, retrace the steps made to find where the team's reasoning went wrong: Why didn't the change have the intended effect? The intent is not to place blame, but a scientific approach to develop understanding and allow for a revised action plan that will be successful.

Tools for data analysis

The *Memory Jogger Plus+* (Brassard 1989) provides a good basic reference for the use of the tools listed below. They are discussed in greater depth in quality-focused texts and books.

The *Affinity Diagram* (Figure 6.1) provides a means to collect very detailed information on a particular issue and then group that information so as to have a better understanding both of the parts and of the whole. For instance, suppose a group were trying to understand the problems of providing technical support for equipment sold by their company. A team of employees involved with customer service would brainstorm for all of the possible causes of problems, such as customers calling in and waiting too long on hold, and warranty replacements delayed because the customer data base was inaccurate, etc. These possible causes afford a detailed representation of the challenges faced by the customer service department. These causes are written on small cards or self-adhesive notes and laid out on a table or stuck to the wall. Silently, team members move cards into what they see as logical groupings. The same card could be moved several times by different team members. After this grouping process is completed, a summary card is written for each group. In this example, these groups could be phone system, computer support system, etc.

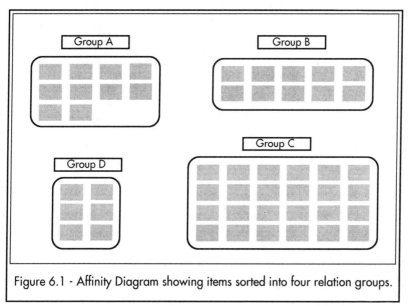

Figure 6.1 - Affinity Diagram showing items sorted into four relation groups.

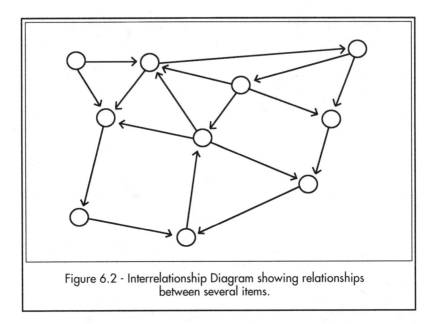

Figure 6.2 - Interrelationship Diagram showing relationships between several items.

The *Interrelationship Diagram* (Figure 6.2) builds on the Affinity Diagram by showing causal or hierarchical relationships between the several elements identified. Each element is written in a box, and lines are drawn between elements to show a cause and effect relationship.

The *Tree Diagram* (Figure 6.3) is used to break a task into its components. One begins with the general task to be accomplished,

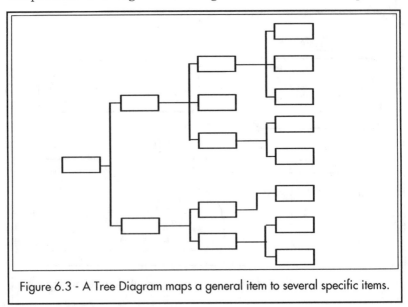

Figure 6.3 - A Tree Diagram maps a general item to several specific items.

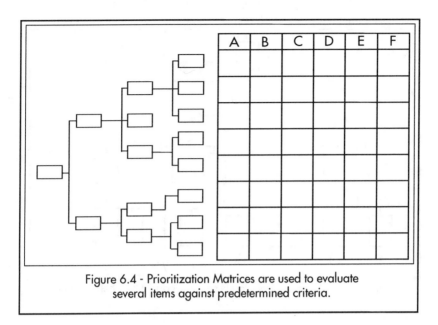

Figure 6.4 - Prioritization Matrices are used to evaluate
several items against predetermined criteria.

which is written in the left-most column of the paper. In the next
column the components of this task are listed. The subcomponents for
each of these tasks are placed in the next column. When complete, a list
of detailed tasks is given in the right-most column. This is a useful
process in planning the implementation of a solution and in developing
the data necessary for an Activity Network Diagram.

	A	B	C	D	E	F	G
1							
2							
3							
4							
5							
6							
7							
8							

Figure 6.5 - A Matrix Diagram can be used to allocate resources
or prioritize items based on predetermined objectives.

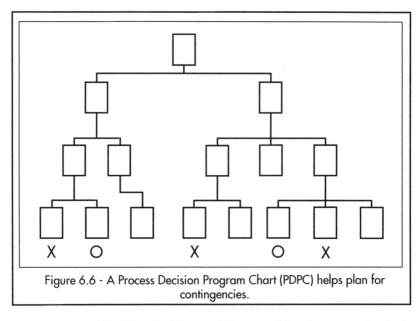

Figure 6.6 - A Process Decision Program Chart (PDPC) helps plan for contingencies.

Prioritization Matrices (Figure 6.4) are useful tools to quantitatively prioritize several options based on given criteria. There are various methods for implementing this tool, but they have the great advantage of obliging the analysts to make a critical and reasoned approach to decision making.

A *Matrix Diagram* (Figure 6.5) helps users understand and define

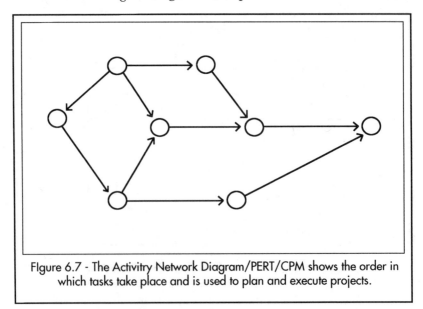

Flgure 6.7 - The Activitry Network Diagram/PERT/CPM shows the order in which tasks take place and is used to plan and execute projects.

the relationships between several items on a list. These diagrams can be two-, three-, or four-dimensional and can be used in allocating resources, evaluation of options, and are also a part of the Quality Function Deployment (QFD) process.

A *Process Decision Program Chart* (PDPC) (Figure 6.6) offers a means to plan for all contingencies in the implementation of a solution. Through theorizing and experimenting, possible outcomes are identified and optimal responses formulated.

An *Activity Network Diagram* (Figure 6.7) is a project management tool that is based on PERT and the Critical Path Method (CPM). These charts are variations of the same basic method and graphically map the tasks that must be completed to finish a project. The two general parameters are precedence, which describes the task or tasks that must be started or completed before work on a subsequent task can begin, and duration, or the time required to complete a task. Using this information, tasks can be managed so that the project is completed in the shortest possible time with the optimal use of resources. Information on PERT and CPM can also be found in any Operations Research text.

The *Flow Chart* (Figure 6.8) gives a visual description of a process. The two major elements of the chart are processes, short descriptions of which are written inside rectangles, and decisions, which are described in diamond-shaped figures. Arrows connect the figures to show the flow of the process. The decision diamonds have two arrows leading to other boxes, one for a yes response, and the other for no. Flow charts are

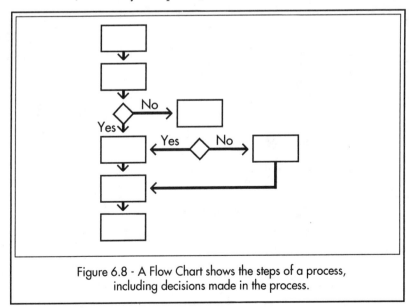

Figure 6.8 - A Flow Chart shows the steps of a process, including decisions made in the process.

Defect	Shift			Total
	Graveyard	Day	Afternoon	
Paperwork error	⦀⦀⦀ ⦀⦀⦀	⦀⦀⦀ ⦀	⦀⦀⦀⦀	18
Product misrouted	⦀⦀⦀⦀	⦀⦀	⦀⦀⦀ ⦀	12
Wrong product shipped	⦀⦀	⦀⦀⦀⦀	⦀⦀	8
Total	14	12	12	38

Figure 6.9 - The Check Sheet is a basic tool for gathering information.

an excellent means of getting agreement on what happens in a process and where problems exist in the process.

The *Check Sheet* (Figure 6.9) is a basic tool for gathering information. Time frames are entered across the top of the chart and various outcomes of interest are listed along the left side of the chart. Each time an outcome occurs, a hash mark is drawn under the time it

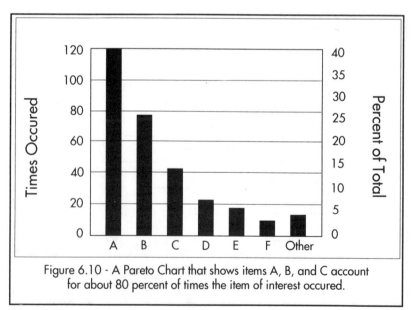

Figure 6.10 - A Pareto Chart that shows items A, B, and C account for about 80 percent of times the item of interest occured.

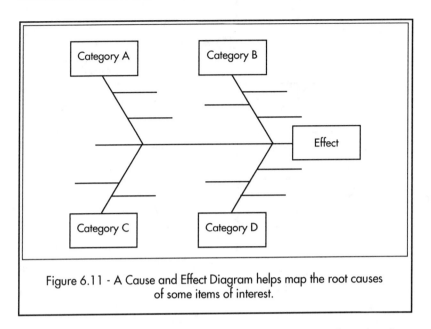

Figure 6.11 - A Cause and Effect Diagram helps map the root causes of some items of interest.

happened. This will show the most common problems and at what time they occur.

A *Pareto Chart* (Figure 6.10) is an analysis that orders elements by the impact they have on the situation being studied. It is based on the principal that the majority of problems come from relatively few causes. Each item is shown on a bar graph, with the magnitude of the problem being the length of the bar. The longest bar is shown on the left of the chart, with each successively smaller bar to its right.

The *Cause and Effect Diagram* (Figure 6.11) is a very useful analytical tool used to enumerate and display the causes for some outcome of interest. The outcome is written on the right side of the paper. Conditions or activities that cause this outcome are written on several spokes that feed into the outcome.

The *Run Chart, Histogram, Scatter Diagram, Control Chart,* and *Process Capability Chart* are statistical tools that can be extremely useful. The use and mechanics of these tools are presented in any of a number of books on statistical process control such as *Introduction to Statistical Process Control,* (Douglas 1985), or *Statistical Quality Control Handbook* (Western Electric 1956, AT&T Technologies 1984).

Brainstorming and the *Nominal Group Technique,* as well as *Multivoting* are team analysis tools that are presented in chapter three.

Force Field Analysis (Figure 6.12) is a graphical means of analyzing people or conditions that will bring change about and those people or conditions that will inhibit desired change. It consists of two

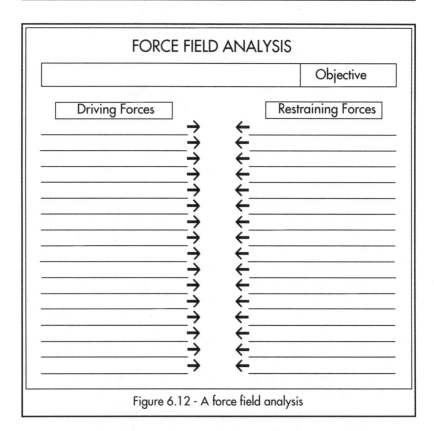

FORCE FIELD ANALYSIS

	Objective

Driving Forces   Restraining Forces

Figure 6.12 - A force field analysis

columns, the left listing enablers and the right giving roadblocks. The analysis is not meant to be one of political intrigue, but rather an honest evaluation of the environment.

Stratification is an analysis tool that provides an opportunity to look for ways to group data that will make the interpretation more obvious. For instance, there may be no apparent pattern of errors in a process, but when the data for each work shift or each operator is analyzed separately, a pattern may emerge. Stratification is a simplified application of statistical experimental design.

Chapter 7

Communication: It's What You Say And How You Say It

Two participants at a recent conference told an amusing story about a communications breakdown. They both worked for the same company, one at a U.S. plant, and the other at a plant in Canada. An employee at the Canadian facility called the U.S. warehouse. The Canadian had an urgent need for a part and asked the American to send it up overnight express. The Canadian pronounced the requested part as deckle (rhymes with heckle). The American didn't understand what the Canadian had asked for, and asked her to repeat it. The Canadian repeated her request for 'deckles.'

The American was becoming very confused. Both plants built similar products, so most of the parts were common to both production lines. The American was experienced and thought he was familiar with all the warehouse parts. It must have been a poor phone connection. Maybe the background noise. So he asked her to repeat the part.

"'Deckles.'"

"'Deckles?'" he asked.

"Yes, 'deckles', and we need them right away or we won't be able to make shipments."

No 'deckles,' no shipments, thought the American, altogether bewildered, and then asked, "Have we sent you 'deckles' before?"

"Yes, of course," replied the confused and irritated Canadian, who began to wonder whether the American was being rude or just playing a joke.

"I'll be honest. I'm sorry, but I have no idea what a 'deckle' is," said the American. "Can you describe it to me?"

Still doubtful, the Canadian explained, "It is removed from a paper backing and applied to the sides of the product. It shows the company name and logo. It gives the machine a sportier appearance. 'Deckles.'"

The light began to get brighter for the American. "Do you mean decals?"

"Of course, 'deckles,' and we need them right away."

It seems not everyone pronounces 'decal' the way Americans do.

There is no end of stories, humorous and tragic, in which lives did not progress as anticipated because of a great disparity between what one person meant and the other understood. These stories all highlight the cardinal rule of communications: Be as careful and thorough as you want. *You can only control what you mean. You can't control what is understood.* We can do our best to influence what our audience understands, but there can be no guarantee of control.

Communication has one goal in mind, namely, convincing members of your audience to take a desired action.

We pointed out in the introduction that an engineer must be technically competent, productive, and effective. To a great extent, communication skills determine how effective we are. Perhaps there was a time when an engineer didn't need exceptional communication skills. Ninety percent of their work was cerebral, not written or verbal. In an "over-the-wall" environment, an engineer's means of output is systematically formatted: drawings, engineering change orders, work instructions, etc., and input is limited.

Communication skills

Communication skills are critical. Few positions exist now in which communication skills are not absolutely essential. The vast majority of enterprises are listening to the voice of the customer and designing for production (more precisely, designing *with* production personnel). Successful companies are developing close alliances between their internal design and production staffs as well as their outside suppliers and end users. Inter-functional teams are becoming more and more responsible for the important improvements in companies. Engineers must adapt to the ever-changing environment of today's business.

Consider the changes that have taken place at Ford Motor Company. Imagine you are an engineer at Ford in the years before World War II. David Halberstam (1986) describes the environment as an extremely confused and disoriented bureaucracy. Henry Ford, after several successful years as a manufacturing pioneer, spent the next several years destroying the progress he had made. He nearly brought the company to ruin by the time the war began. His managers were car men, and came up through the ranks. Ford's method for maintaining control was to pit one top manager against another, ensuring that neither of them gained too much power. He managed by "gut feel," and purposely kept poor financial records to protect himself from the Internal Revenue Service.

In this environment, engineers had to be very adept at staying out of harm's way. They had to second guess what might be viewed favorably by upper management, whose views about cars began to reflect car buyers' preferences less and less. When asked by some of his dealers who were concerned about competing with the growing GM enterprise, if the Model T would be available in different colors, Ford gave his well-known reply, "You can have them any color you want boys, as long as they're black." Most important, loyalty to leaders and conformance to the "party line" was valued over radical thinking.

There are companies today that are similar in their organization. If working for such a company, what would you do to be effective in making a positive contribution?

John Byrne (1993) details the dramatic change in the culture at Ford during the boom years following the war. At the insistence of the Board of Directors, Ford resigned in the latter part of 1945. Henry Ford, Henry's grandson, was made president. The younger Henry was not like his grandfather. One thing was clear to him, Ford Motor Company must somehow be brought under control.

He was quite young and did not have a lot of experience in the car business. Charles Thornton, a young army officer who led an exceptionally bright group of other young officers in dealing with logistics during the war, approached Ford with an offer to bring the control to Ford Motor that his group had brought to the war. This cadre included such notables as Robert McNamara and Arjay Miller. They became known as the "Whiz Kids."

Ford accepted his offer. No longer were decisions made off-the-cuff. It was management by numbers, financial numbers in particular. Unfortunately, the engineers and production personnel were not well equipped to analyze their plans in a financial light. They were even less prepared to present their plans to decision makers who asked endless questions about cost-effectiveness. The engineers didn't understand their audience, and the audience had a very low tolerance for the engineer's inability to talk in terms of finance. In this environment, conformance to sound financial principles was valued over loyalty to leaders. Also in this environment, the Pinto was born (one of Ford's biggest failures) and the Japanese made astounding gains in the U.S. auto market.

In this changed environment, how could an engineer be effective? Perhaps if the engineers had been more adept at communicating in terms of income and balance sheets, Ford would not have lost its competitive edge over foreign car makers. Besides the obvious failures of the Edsel and Pinto models, there was at least one exception: Lee Iacocca's marketing of the Mustang. Iacocca was able to speak both the language of the engineer, assisted by Donald Frey (a manager in the engineering group), as well as the language of the finance-obsessed managers. The Mustang was an immediate high-profit success.

Ford's environment, as well as that of every other organization, continues to change. To compete in today's market, loyalty from customers is highly valued. Being truly responsive to customers requires a very open environment. Learning and teaching has become an important part of everyone's work. The following discussion presents elements that are critical for successful communication in this environment.

Elements of successful communication

To present ideas successfully, carefully consider the needs and nature of the audience to determine content and format. Using a customer-focused approach, design communication like you would design a new product. The question is not only, "What do I need or want to tell them?" but also, "What will they need and want to know?" A significant portion of the effort put into communication should be spent on this preparation.

First, you should determine the audience and your intended purpose for communicating with them, including:

- Who are they (their area of responsibility)? What is their expertise?
- What are their competencies and blind spots?
- What are their concerns?
- What actions should they take based on your communications?

Then, select material to suit the audience and meet your purpose, including data and anecdotal information. It is not the audience's fault if they don't get it or if they are not convinced of your recommendation.

Understand the subject from as many different angles as you can imagine. Enlist the help of others to see new angles. Gain a working understanding of all this information. The next step is difficult: Glean the most essential information from all this detail. It is a necessary and often frustrating fact that generally only a small fraction of the information you have prepared will be passed on to your audience. Do this refining with a strong focus on the customer.

Next, select a forum. There are several modes of communication, and all of them should become a part of your style. Everyone has areas of comfort. For instance, some people are much more comfortable with a face-to-face meeting than in telephone conversation. Build and rely on strengths and recognize and overcome weaknesses.

Finally, elect a format that is most conducive to your audience, material, and forum. Formats could include overheads or slides, memos, speaking notes, and reports. Consider the level of detail that will be appropriate. Make good use of graphics and other visuals to make your point clear.

All communications can take on varying degrees of formality. Successful presentations, for instance, can range from a carefully prepared slide show to a back-of-the-napkin impromptu discussion. Again, pursue all preparation with your goal in mind: convincing your audience to take some action. Choose the format and level of formality that will most likely accomplish your purpose.

Means of communication

The common means of communication are: face-to-face, oral only, and written. Each has its own strengths and weaknesses.

Face-to-face

Face-to-face communications offer a significant advantage: Communications can be monitored and modified in real time, ensuring feedback. It is much easier to detect and respond to nuances such as intonation and facial expressions.

There is a disadvantage, however. The dialog is harder to organize and control, and it is harder for the recipient to retain the information. Communication has not been successful if it is not remembered.

Face-to-face communication can either be one-on-one or with a group of people. The one-on-one is useful for communications that are narrow in their appeal or for winning support of an idea ("greasing the skids"). It is also useful for sensitive information or to avoid group-think (a counterproductive condition in which members don't express their own thoughts but either consciously or subconsciously try to anticipate the input that will be favored by the group).

Group settings are appropriate for spreading information, brainstorming, or building consensus. However, one must be very cautious in presenting information that could be damaging or threatening to a member of the group in front of their peers. At some times it may be appropriate, maybe when all other avenues have been exhausted, but it is generally counterproductive. For example, remember your student career. How successful would you have been trying to negotiate a better grade while in the class setting, with the other students listening? Success would be much more likely in a meeting alone with the professor.

Oral only (phone, etc.)

The advantages of oral communication are the same as face-to-face. Phone calls are generally easier to arrange than in-person meetings.

However, group interaction is greatly limited in conference calls. Also, the feedback provided through facial expressions is not available.

Written

Written communication can be more detailed, complete, and concise than oral communication. Documents can be carefully constructed, reducing the likelihood of misunderstanding.

However, it is generally more time consuming to write a document than to prepare for verbal communication. One is also less assured that the communication will actually take place, since the document may or may not be read.

Attributes of successful communication

Completeness

Everything that the receiver needs and wants to know should be included. We mentioned that generally more information will be available to the audience than you will actually be able to give them. Since you cannot be certain which questions will be asked, it is best to prepare for all imaginable questions and directions the discussion may take.

Accessiblility

Information should be easily obtainable by the receiver. It is likely that you will not have very much time to control the attention of your audience. Make it as easy as possible for them to discern the points of your communication.

When a document arrives at someone's desk, one of three things will happen. The person will read the document and act on it; the person will throw it away immediately; or the person will put it in a pile of things to be dealt with later. This decision is typically made in less than five seconds! Your point and its importance must be readily available to the reader.

It should be recognized that most people have this pile of material to be dealt with later. (Check your own or your boss' desk). And, as much as we hate to admit it, it usually just postpones the decision to throw it away.

Try the following experiment. Take 30 seconds to read the memo in Figure 7.1 as thoroughly as that time permits. Then make a note of all the information you remember. When you have finished your list, read the revised memo in Figure 7.2, again allowing only 30 seconds, and again list important details. Compare your lists. Given the same amount of time, most people get much more information out of the second memo. Both documents have the same content, but the second memo provides the information much more readily, both at the initial reading and on review.

Brevity

Because you will generally have a very limited time to make your point, eliminate all that is not essential. This must be done with great care and consideration to ensure that completeness is not sacrificed.

Kennecott Corporation's Bingham Copper Mine is located near Salt Lake City, Utah, and is purportedly the largest open-pit copper mine in the world. Copper is mined there by brute force. For every 3 tons of material removed, approximately 10 pounds of copper are obtained. (The ore has about 0.6 of one percent copper content).

Business communications should follow a similar process of refining. Begin with the bulk, lots of ideas and facts. Work and rework, combine and dissect, include and eliminate until you are left with the barest essentials. Your work must have the substance to be understandable and convincing, but free of fluff. Give your audience the copper, silver, and gold, not tons of rock.

Case in point

A controversial figure of recent times who demonstrated a mastery of communication skills is Oliver North. Here is a man who went before the United States Senate (and through television, a large portion of the nation), and told them he had lied, broken laws, and shredded evidence. Yet he was elevated to hero status by a surprisingly large group of people.

Although it is very tempting to get caught up in the merits of the case itself, let's divorce ourselves from the content of this example and evaluate the case from a communication point of view. Our first point in this chapter was preparation. Colonel North was definitely well prepared. He knew his audience, which wasn't necessarily the senators, but instead, the American people. His purpose, then, was to legitimize

TO: All Managers and Supervisors
FROM: Total Quality Implementation Team
DATE: January 1
RE: Total Quality

As you know, our company will begin the implementation of total quality management concepts and practices with training sessions for all managers and supervisors. This training will present an overview of basic principles of quality management. The first session will include the basics of the elementary quality tools: scatter diagrams, histograms, control and run charts, Pareto charts, cause and effect diagrams, flowcharts, work-flow diagrams, check sheets, brainstorming, and Nominal Group Technique. The first four tools are based on statistics. The next four are used to analyze processes and identify problems. The last two are useful in formulating solutions.

The second session will address planning, beginning with the selection of projects on which your area will work. This will include a discussion of critical items to consider and projects to avoid, followed by the means by which team leaders and members should be selected and the great importance this has on the eventual success of the team. The important role of the team sponsor will be examined. Projects will be successful when team sponsors suppress the natural inclination to be involved in every aspect of the teams' operations. The team sponsor is needed to assist the team leaders in their work, leaving them considerable latitude in the day-to-day operations of the group. It will be important that sponsors recognize things that will set up unnecessary organizational roadblocks to the teams and take the responsibility to remove these roadblocks. Finally, the importance of providing recognition, and, where appropriate, rewards for teams as they meet their goals will be reviewed.

Classes will begin in February and continue through the end of March. Session One will be taught on Tuesday, and Session Two on the following Thursday. It is important that you attend both sessions in the same week. Each class will begin at 8:00 am and finish no later than 4:00 p.m. Please select the week in which you will take the classes and call the Training Coordinator with your choice no later than January 20. You can prepare for the course by reading the accompanying refresher on total quality.

Figure 7.1 Original Memo.

TO: All Supervisors
FROM: Total Quality Implementation Team
DATE: January 1
RE: Implementation of Total Quality Concepts and Practices
          Introductory Training, Scheduling and Description

You will need to attend one of the two-day training sessions
scheduled for all managers and supervisors during the months of
February and March as part of the implementation of total quality
management concepts and practices in our company. **Please select
one of the training groups listed below and call the Training
Coordinator with your choice no later than January 20.**

Training Schedule (Select a two-day group, classes run 8:00- 4:00)

Mar. 7,9	Feb. 8,1
Mar. 14,16	Feb. 15,17
Mar. 21,23	Feb. 22,24
Mar. 28,30	Feb. 28. Mar. 2

Course Outline
> ***Session One - Elementary Quality Tools***
> > **Statistics**
> > > Scatter diagrams, Histograms, Control
> > > charts, Run charts
> > **Process Analysis and Problem Identification**
> > > Cause and effect diagrams, Flowcharts,
> > > Work-flow diagrams, Check sheets,
> > > Brainstorming, Nominal Group
> > > Technique
> ***Session Two - Planning***
> > **Selecting Projects**
> > > Critical items to consider, Projects to
> > > avoid
> > **Selecting Teams**
> > > Team leaders, Team members, Impor-
> > > tance of careful selection
> > **Role of the Team Sponsor**
> > > Facilitate, don't operate
> > > Recognizing and removing roadblocks
> > > Recognizing and rewarding success

Preparation
> Read the accompanying refresher on total quality
> Call the Training Coordinator by Jan. 20 with
> your schedule choice

Figure 7.2 Memo revised for accessibility.

his actions in the court of public opinion. He didn't get to choose his forum, but knowing how a senate hearing worked, he planned his assault with great care and attention to detail.

Many engineers will find themselves in a similar business situation. Often the forum is set and beyond their control and more than a little intimidating. For example, one may be given twenty minutes in the executive staff meeting to present the results of a project or assignment. Several month's work comes down to a twenty-minute presentation. Of course, your proposal will not be the only item considered at the meeting. You may have twenty minutes to convince them they should proceed with your recommendation rather than other projects competing for the company's resources. Usually in these meetings the agenda is quite flexible. If things go well and you capture their attention they will ask questions, giving you more time and greater opportunity to make your case. Similarly, if your preparation is lacking, your time may be cut short.

North attended the hearings in full United States Marine Corps uniform, not the common suit he otherwise wore in his work at the White House. He presented himself to the committee and the public as a well-decorated Marine with a chest full of ribbons. He wanted everyone to know he was no ordinary White House staffer.

It is important to pay attention to appearance in order to meet the expectations of your audience. When visiting a job site with a construction crew, work clothes may suggest competence and practicality. Work clothes would not likely be appropriate in the board room, though. The way an engineer looks and acts when discussing the operation of a bulldozer with a foreman is altogether different than when asking the executive committee to fund its development.

Another lesson engineers can learn from the Oliver North case is speaking with confidence during presentations. Colonel North answered the questions put to him with confidence and conviction. Successful engineers study and research issues regarding their projects or proposals until they have mastered them. Based on this preparation they speak with confidence, whether in a one-on-one conversation with a colleague or during a presentation to senior management. Conveying a sense of confidence and keeping one's composure under pressure can make the difference between success or having to wait for another opportunity, as evidenced by President Kennedy's success over Richard Nixon. Kennedy's victory in the 1960 presidential election has been attributed largely to his talent as a communicator during the nationally televised debate.

Conclusion

Among the six essential career skills we have covered in this book, communication has the largest impact on all the other skills. A shortcoming in this area can potentially reduce one's ability to achieve success in general. Poor communication can, and probably will, result in poor customer satisfaction. It can have a disastrous effect on a team process. Inability to communicate a clear picture on managing technology will result in its poor management. The story of dealing with the "Whiz Kids" stresses the importance of financial skills as well as the ability to express that financial understanding to senior management. Lastly, communication problems can seriously jeopardize the accuracy of data gathering and as a result reduce the effectiveness of the engineer with the best data analysis skills. With such a broad impact, it is critical that engineers continuously evaluate and update their communication skills. Especially with this skill, the best way to improve is to watch others, take appropriate courses, research, and practice. The key role that effective communication plays in the success of one's career cannot be overstated. No less in business than in politics, those who are able communicators are those who succeed.

A Final Note

We have provided an overview of the first six variables in the formula for success. We wish you the best as you enhance your skills in each of these areas as well as identify and develop the rest of your skill set ($V_7...V_n$) needed for a rewarding and fruitful career.

V_7 may be an industry-specific skill. V_8 may have to do with your employer. V_9 may depend on where you live or where you want to work, such as skills in a second language. As you search for these variables in the formula and determine the weight associated with each one, as well as the relationship between them, don't lose sight of the big picture. Remember: They probably won't say they are looking for holistic engineers, but they are!

Bibliography

Abdu'l-Baha. 1982. *Selections from the writings of Abdu'l-Baha.* Haifa: Bahai World Centre.

Argyris, Chris. 1985. *Strategy change and defensive routines.* Boston: Pitman Boston.

Barkume, Megan. 1992. Computers: Instruments of change. *Occupational Outlook Quarterly.* 36(4). Winter, 2-3.

Batz, Jeannette. 1993. *Half life: What we give up to work.* St. Louis: Virginia Publishing.

Brandreth, Gyles. 1984. *Numberplay.* New York: Rawson Associates.

Brassard, Michael. 1989. *The memory jogger plus+.* Metheun, Massachusetts: Goal/QPC.

Byrne, John A. 1993. *The whiz kids.* New York: Doubleday.

Cook, William J. 1994. The end of the plain plane. *U.S. News & World Report.* 11 April, 43-47.

Davidow, William H. and Bro Uttal. 1989. *Total customer service.* New York: Harper Perrenial.

Davidson, Sticknay and Weil. 1988. *Financial accounting: An introduction to concepts, methods and uses* (5th ed.). Orlando: Harcourt Brace Jovanovich.

Dornheim, Michael A. 1991. Computerized design system allows Boeing to skip building 777. *Aviation Week & Space Technology.* 3 June, 50-51.

Fisher, Christy. 1991. Wal-Mart's way. *Advertising Age,* 18 February, 3, 48.

Grant and Levenworth. 1980. *Statistical quality control.* McGraw-Hill.

Halberstam, David. 1986. *The reckoning.* New York: William Morrow and Co.

Hronec, Steven M. 1993. *Vital signs.* New York: AMACOM.

Katzenbach, Jon R., and Douglas K. Smith, The Wisdom of Teams, *Harper Business 1994.*

Montgomery, Douglas C. 1985. *Introduction to statistical quality control.* Somerset, New Jersey: John Wiley & Sons.

National Transportation Safety Board. 1979. Aircraft Accident Report #NTSB-AAR-79-17. 21 Dec.

Neidert, David L. 1994. Leading people. *Executive Excellence.* 11(2) Feb.

New Scientist. 1993. Commentary. 10 April.

O'Lone, Richard G. 1992. Final assembly of 777 nears. *Aviation Week & Space Technology.* 12 Oct., 48-50.

Proctor, Paul. 1993. 777 detailed design nearly complete. *Aviation Week & Space Technology.* 19 April, 35.

Senge, Peter. 1990. *The fifth discipline.* New York: Doubleday/Currency.

Sewell, Carl, and Paul B. Brown. 1990. *Customers for life.* New York: Doubleday.

Stover, Dawn. 1994. The newest way to fly.. *Popular Science.* June, 78,79,104.

Toronto Sun. 1983. 25 April.

Vonnegut, Kurt. 1980. *Player piano.* New York: Dell Publishing.

Wheatley, Meg. 1992. *Leadership and the new science.* San Francisco: Berrett-Koehler.

Glossary

activity network diagram. A project management tool that is based on PERT and the Critical Path Method (CPM). These charts are variations of the same basic method and graphically map the tasks that must be completed to finish a project. The two general parameters are precedence and duration.

affinity diagram. A technique used to collect very detailed information on a particular issue and then group that information so as to have a better understanding both of the parts and of the whole.

balance sheet. A financial statement which provides information about the assets of an enterprise and how they are financed at a particular point in time. It is called a balance sheet because the financing must be equal to the assets.

brainstorming. A method used by a group of people to quickly generate ideas within a given time frame.

capital budget. A budget or plan of proposed acquisitions and replacements of long-term assets and their financing. A capital budget is developed using a variety of capital budgeting techniques such as the discount cash flow method.

cash flow statement. A statement showing from what sources cash has come into the business and on what the cash has been spent.

cause and effect diagram. An analytical tool used to enumerate and display the causes fro some outcome of interest. The outcome is written on the right side of the paper. Conditions or activities that cause this outcome are written on several spokes that feed into the outcome.

check sheet. A basic tool for gathering information. Time frames are entered across the top of the chart and various outcomes of interest are listed along the left side of the chart. Each time an outcome occurs, a hash mark is drawn under the time it happened. This shows the most common problems and at what time they occur.

chunking. Implementing change one step at a time.

cross functional teams. Teams whose members are from several work units that interface with one another. These teams are particularly useful when work units are dependent on one another for materials, information, and the like.

defensive routines. Habits that are developed to protect oneself when conflicts arise and the people are unwilling to discuss the issues and alternatives.

depreciation. The procedure of spreading out the acquisition cost of fixed assets (such as machinery and equipment) to each of the time periods in which they are utilized.

efficiency. The reduction of inputs and the maximization of outputs

empowerment. Defining an individual's job as widely as possible and giving each individual the authority and accountability to make the decisions that affect his or her work.

flow chart. A pictorial representation showing all of the steps of a process. Flow charts can be used to show how various steps in a process are related to each other.

force field analysis. A technique that displays the Driving (Positive) and Restraining (Negative) forces surrounding any change. It is displayed in a balance sheet format.

Gantt chart. A type of planning chart especially designed to show graphically the relationship between planned performance and actual performance.

holistic engineer. An engineer who combines several interdependent skills and this combination is greater then the sum of the parts. Holistic implies an understanding of the big picture, an appreciation for all variables and an avoidance of sub-optimization

income statement. Reflects the results of the operation of an enterprise for a period of time.

interrelationship diagram. Builds on the affinity diagram by showing causal of hierarchical relationships between the several elements identified. Each element is written in a box, and lines are drawn between elements to show a caused and effect relationship.

matrix diagram. A tool used to help define the relationships between several items on a list. These diagrams can be two-, three-, or four-dimensional and can be used in allocating resources, evaluation of options, and are also part of the Quality Function Deployment (QFD) process.

multivoting. A means to condense a large list of ideas down to a smaller list of the most important.

nominal group technique. A weighted ranking technique that allows a team to prioritize a large number of issues without creating "winners" and "losers."

operating budget. A budget that embraces the impacts of operating decisions. It contains forecasts of sales, net income, the cost of goods sold, selling and administrative expenses and other expenses.

Pareto chart. A special form of vertical bar graph which helps to determine which problems to solve in what order.

PERT (program evaluation and review technique) chart. A probabilistic technique used mostly by government agencies, for calculating the most likely durations for network activities.

prioritization matrices. Tools used to quantitatively prioritize several options based on given criteria.

process decision program chart. A technique that offers a means to plan for all contingencies in the implementation of a solution. Through theorizing and experimenting, possible outcomes are identified and optimal responses formulated.

run chart. Employed to represent data visually. They are used to monitor a process to see whether or not the long range average is changing.

scatter diagram. Used to study the possible relationship between one variable and another. The scatter diagram is used to test for possible cause and effect relationships. It cannot prove that one variable causes the other, but it does make it clear whether a relationship exists and the strength of that relationship.

stratification. An analysis tool that provides an opportunity to look for ways to group data that will make the interpretation more obvious.

sunk costs. A cost which, since it occurred in the past, has no relevance with respect to estimates of future receipts or disbursements. This concept implies that, since a past outlay is the same regardless of the alternative selected, it should not influence a new choice among alternatives.

tree diagram. Technique used to break a task into its components.

Index

T

U, V

W, X, Y

About the Authors

Shahab Saeed is a registered professional engineer with a master's degree in business administration from the University of Utah. He is listed in *Who's Who in Engineering*. He has extensive experience in the energy industry with Questar Corporation and Mountain Fuel Supply in a variety of management positions. Currently he serves as Assistant to the President for Continuous Improvement.

Keith Johnson has worked in electronics manufacturing and now works in the continuous improvement group of Mountain Fuel Supply, a local natural gas distribution company. He is a registered professional engineering and ASQC certified quality engineer. He holds a master's degree in business administration from Brigham Young University.

About the Series

More and more people entering today's work force are being asked to perform multiple tasks that may include those once limited to traditional industrial engineers. The Engineers in Business Series is about understanding and applying a combination of business skills and industrial engineering principles. It was developed to meet the needs of students, new industrial engineers, and individuals without engineering degrees interested in learning about the different aspects of the industrial engineering profession.

About IIE

Founded in 1948, the Institute of Industrial Engineers (IIE) is comprised of more than 25,000 members throughout the United States and 89 other countries. IIE is the only international, nonprofit professional society dedicated to advancing the technical and managerial excellence of industrial engineers and all individuals involved in improving overall quality and productivity. IIE is committed to providing timely information about the profession to its membership, to professionals who practice industrial engineering skills, and to the general public.

IIE provides continuing education opportunities to members to keep them current on the latest technologies and systems that contribute to career advancement. The Institute provides products and services to aid in tis endeavor, including professional magazines, periodicals, books, conferences and seminars. IIE is constantly working to be the best available resource for information about the industrial engineering profession.

For more information about membership in IIE, please contact IIE Member and Customer Service at (800) 494-0460 or (770) 449-0460.